COMMON WAVEFORM ANALYSIS:
A New And Practical Generalization of Fourier Analysis

The Kluwer International Series on
ASIAN STUDIES IN COMPUTER AND INFORMATION SCIENCE

Series Editor

Kai-Yuan Cai

Beijing University of Aeronautics and Astronautics
Beijing, CHINA

Also in the Series:

AUTOMATED BIOMETRICS: ***Technologies and Systems***, *by David D. Zhang*; ISBN: 0-7923-7856-3

FUZZY LOGIC AND SOFT COMPUTING, *by Guoqing Chen, Mingsheng Ying, Kai-Yuan Cai*; ISBN: 0-7923-8650-7

INTELLIGENT BUILDING SYSTEMS, *by Albert Ting-pat So, Wai Lok Chan*; ISBN: 0-7923-8491-1

PERFORMANCE EVALUATION, PREDICTION AND VISUALIZATION OF PARALLEL SYSTEMS *by Xingfu Wu*; ISBN: 0-7923-8462-8

ROBUST MODEL-BASED FAULT DIAGNOSIS FOR DYNAMIC SYSTEMS *by Jie Chen and Ron J. Patton;* ISBN: 0-7923-8411-3

FUZZY LOGIC IN DATA MODELING: ***Semantics, Constraints, and Database Design*** *by Guoqing Chen;* ISBN: 0-7923-8253-6

SOFTWARE DEFECT AND OPERATIONAL PROFILE MODELING *by Kai-Yuan Cai;* ISBN: 0-7923-8259-5

COMMON WAVEFORM ANALYSIS:
A New And Practical Generalization of Fourier Analysis

by

Y. Wei
and
Q. Zhang
Beijing University of Aeronautics and Astronautics
Beijing

SPRINGER SCIENCE+BUSINESS MEDIA, LLC

Library of Congress Cataloging-in-Publication Data

Wei, Y. (Yuchuan), 1966-
Common waveform analysis : a new and practical generalization of Fourier analysis / by Y. Wei and Q. Zhang.
p.cm. -- (The Kluwer international series on Asian studies in computer and information science ; 9)
Includes bibliographical references and index.
ISBN 978-1-4613-7052-9 ISBN 978-1-4615-4547-7 (eBook)
DOI 10.1007/978-1-4615-4547-7
1. Fourier analysis. I. Zhang, Q. (Qishan), 1936-II. Title. III. Series.

QA403.5 .W45 2000
515'.2433--dc21

00-041593

Printed on acid-free paper.

SERIES EDITOR'S ACKNOWLEDGMENTS

I am pleased to acknowledge the assistance to the editorial work by Beijing University of Aeronautics and Astronautics and the National Natural Science Foundation of China.

Kai-Yuan Cai
Series Editor
Department of Automatic Control
Beijing University of Aeronautics and Astronautics
Beijing 100083
China

Contents

1 ABC OF NUMBER THEORY 1

1.1 Divisibility 1

1.2 Arithmetical Functions 2

1.3 Dirichlet Multiplication 6

1.4 Dirichlet Series 9

2 SQUARE WAVE ANALYSIS 13

2.1 Square Wave System and its Basic Properties 14

2.2 Biorthogonal Functions and Square Wave Series 22

2.3 Orthogonalization and the Best Approximation 30

2.4 An Example of Applications 37

3 TRIANGULAR WAVE ANALYSIS AND TRAPEZOIDAL WAVE ANALYSIS 41

3.1 WASCMFC Functions and Practical Examples 42

3.2 WASCMFC Function Basis, Biorthogonal Basis and Orthonormalized Basis 49

3.3 Basis and Coordinate Transforms 54

3.4 Discrete Triangular Wave Transform and Trapezoidal Wave Transform 65

4 FREQUENCY ANALYSIS BASED ON GENERAL PERIODIC FUNCTIONDS 75

4.1 A Frequency System in $L^2[-\pi, \pi]$ 76
4.2 A Frequency System in $L^2_{\text{odd}}[-\pi, +\pi]$ 79
4.3 A Complete System in $L^2_{\text{odd}}[-\pi, +\pi]$ 83
4.4 An Unconditional Basis in $L^2_{\text{odd}}[-\pi, +\pi]$ 95
4.5 A Combinative Frequency System in $L^2[-\pi, \pi]$ 101
4.6 A Frequency Transform in $L^2(R)$ 104

5 **MAIN RELATIONS AND BASIC TECHNIQUES** **113**
5.1 Dirichlet Multiplication and a Related Formula. 114
5.2 Relations between Sine Waves and Common Waveforms . 117
5.3 Relations between Two Waveforms 124
5.4 Common Waveform Series 126
5.5 Common Waveform Transform 135
5.6 Discrete Transform for Common Waveform 141
5.7 Techniques of Common Waveform Analysis 145

PREFACE

In this monograph we study common waveform analysis systematically, which is a new and practical generalization of classic Fourier analysis.

About two hundred years ago, the French mathematician J. B. J. Fourier asserted that any function with period 2π can be expressed as a trigonometric series. This great idea has had an important influence upon science and technology. In electronics, Fourier analysis has been playing an important role, and a signal is often considered as a superposition of many sine and cosine functions with different frequencies.

However, with the development of electronic technique, sawtooth wave, square wave, triangular wave, trapezoidal wave and so on become new easily-generated periodic functions, or in electronic terms, common waveforms. Thus the following questions arise naturally:

(1) Can a signal be considered as a superposition of common waveforms with different frequencies?

(2) How to decompose a signal into a series of common waveforms with different frequencies?

(3) How to approximate a signal best with finite common waveforms?

(4) How to find a combination of common waveforms that equals a given signal at N uniform points?

(5) Can common waveforms be used in techniques as well as sine-cosine functions?

In order to answer these practical questions systematically, we generalize classical Fourier analysis to frequency analysis based on common waveforms, shorten as Common Waveform Analysis. Because of the practical importance, we pay special attention to square wave analysis, triangular wave analysis and trapezoidal wave analysis. Associated with Fourier series, Fourier transform, and discrete Fourier transform, we introduce common waveform series, common waveform transform, and discrete common waveform transform, respectively. These results make it possible to represent a signal by a superposition of common waveforms. A lot of techniques based on sine-cosine functions can be translated into those based on common waveforms.

This monograph is divided into 5 chapters.

Since common waveform analysis has a close relation with number theory, Chapter 1 introduces some basic concepts and simple propositions in number theory, which will be used in the following chapters.

Since square wave is the most important common waveform, square wave analysis is considered firstly in Chapter 2. Different from Walsh analysis or Haar analysis, the square wave analysis introduced in this monograph has a simple and close relation with Fourier analysis.

Triangular wave and trapezoidal wave are common waveforms as well. Since triangular wave analysis and trapezoidal wave analysis have similar properties, they are discussed in the same chapter. The relations between triangular wave analysis, trapezoidal wave analysis, and Fourier analysis are presented in Chapter 3. The discrete triangular wave transform and trapezoidal wave transform are introduced as well.

Since electrical waveforms may have various shapes, it is necessary to study frequency analysis based on general periodic functions. In Chapter 4, almost every electrical waveform is discussed as an example. In addition to frequency series based on general periodic functions, the frequency transform based on general periodic functions is introduced and discussed as well.

Functional analysis is too difficult for many readers, so it is necessary to describe the main results of common waveform analysis in an easy way. Independent of other chapters, Chapter 5 is an easy and compressed version of this monograph. The main relations and basic techniques of common waveforms are presented. The authors suggest readers go through this chapter first.

As the contributors of common waveform analysis, we not only give the latest results in this field, but also tell researchers what to do next. All the unsolved theoretical problems are listed in this book. In the technical aspect of common waveform analysis, the step from schematic diagrams to real devices is a systematic engineering. Generally common waveform analysis is a challenge as well as an opportunity for both mathematicians and engineers.

Electrical engineers, mathematicians and students from all over the world including USA, UK, Australia, Germany, Russia, Japan, etc, have been asking us for the reprints of our papers and inquiring about this new method, but we are very sorry that we could not reply them one by one because of our limited time. We wish this book could bring them some help.

Finally we would like to thank the editors of Kluwer Academic Publishers, and those who helped us. This work is supported by the Chinese National Natural Science Foundation and the Chinese Postdoctoral Science Foundation. We are willing to cooperate with the institutions which are interested in this field.

Y. Wei and Q. Zhang
Beijing, China
September 2000

Chapter 1

ABC OF NUMBER THEORY

It is interesting that common waveform analysis has a close relation with number theory. A series of basic concepts in number theory are encountered frequently, such as the Möbius function, the divisor function, the Dirichlet multiplication, the completely mutiplicative function, and so on [1][2]. Therefore we have to introduce some concepts and prove some propositions, which will be used in following chapters.

1.1 Divisibility

$1, 2, 3, \cdots$ are called positive integers. In this chapter, we suppose that the letters d, m, n, k denote positive integers unless otherwise indicated.

Definition. we say d divides n and we write $d \mid n$ if there exists an positive integer m such that $n = md$. We also say that n is a multiple of d, or that d is a divisor (or factor) of n. If d does not divide n, we write $d \nmid n$.

Example. 2 is a factor of 6, i.e., $2 \mid 6$, but 4 is not a factor of 6, i.e.,

$4 \nmid 6$.

Definition. We say that d is a common factor of m and n, if d is a factor of both m and n. We denote the greatest common divisor of m and n by (m, n).

Example. 1 is a common divisor of 4 and 6; 2 is the greatest common divisor of 4 and 6, i.e. $2 = (4, 6)$.

Definition. We say k is a common multiple of m and n if k is a multiple of both m and n. We denote the least common multiple of m and n by [m,n].

Example. 24 is a common multiple of 4 and 6; 12 is the least common multiple of 4 and 6, i.e. $12 = [4, 6]$.

For any two positive integers m and n, we have the following simple relation [1]:

$$mn = [m, n](m, n) \tag{1.1}$$

1.2 Arithmetical Functions

Definition. A function defined on the positive integers is called an arithmetical function.

Example.

1. The zero function 0 is defined by

$$0(n) = 0 \qquad \text{for } n = 1, 2, 3, \cdots. \tag{1.2}$$

2. The identity function I is defined by

$$I(n) = \begin{cases} 1 & n = 1, \\ 0 & n = 2, 3, 4, \cdots. \end{cases} \tag{1.3}$$

3. The function ϵ is defined by

$$\epsilon(n) = 1 \qquad \text{for } n = 1, 2, 3, \cdots. \tag{1.4}$$

4. The function θ is defined by

$$\theta(n) = (-1)^{n+1} \qquad \text{for } n = 1, 2, 3, \cdots. \tag{1.5}$$

There are two arithmetical functions which are frequently used in this book. One is the Möbius function $\mu(n)$, and the other is the divisor function $d(n)$ [1].

1. The Möbius function μ

The Möbius function μ is defined by

$$\mu(n) = \begin{cases} 1, & \text{if } n = 1, \\ (-1)^r, & \text{if } n \text{ is the product of } r \text{ distinct primes}, \\ 0, & \text{if } n \text{ is divisible by a prime square.} \end{cases} \tag{1.6}$$

Notice that $\mu(n)$ takes three values $1, -1$, and 0 only. The first ten values are $\mu(1) = 1,\ \mu(2) = -1,\ \mu(3) = -1,\ \mu(4) = 0,\ \mu(5) = -1,\ \mu(6) = 1,\ \mu(7) = -1,\ \mu(8) = 0,\ \mu(9) = 0,\ \mu(10) = 1.$

The Möbius function μ has a simple property that

$$\sum_{d|n} \mu(d) = \begin{cases} 1 & n = 1, \\ 0 & n = 2, 3, 4, \cdots, \end{cases} \tag{1.7}$$

where d runs over the positive factors of n including 1 and n.

For example, for $n = 6$, we have

$$\sum_{d|6} \mu(d) = \mu(1) + \mu(2) + \mu(3) + \mu(6) = 0.$$

2. The divisor function $d(n)$

The function d(n) is defined as the number of positive divisors of an integer n. For example, $d(1) = 1,\ d(2) = 2,\ d(3) = 2,\ d(4) = 3$, and so on.

The divisor function $d(n)$ has the well-known properties that

$$d(mn) \le d(m)d(n), \tag{1.8}$$

and that

$$d(n) = O(n^{\varepsilon}), \tag{1.9}$$

where m, n are any two positive integers, and ε is any positive real number.

Definition. An arithmetical function f is called multiplicative [2] if f is not identically zero and if

$$f(mn) = f(m)f(n) \qquad \text{whenever } (m,n) = 1. \tag{1.10}$$

A multiplicative function is called completely multiplicative if we have

$$f(mn) = f(m)f(n) \qquad \text{for all } m,n. \tag{1.11}$$

Obviously, if $f(n)$ is multiplicative, then $f(1) = 1$

Example. The function $\theta(n)$, the Möbius function $\mu(n)$ and the divisor function $d(n)$ are multiplicative but not completely multiplicative.

Example. The function $I(n)$ and $\epsilon(n)$ are completely multiplicative.

Example. The function $f(n) = \frac{1}{n^\alpha}$ (where α is a real number) is completely multiplicative.

Example. By the definition one can verify that the following four functions are completely multiplicative:

1. the function χ_1 defined by

$$\chi_1(n) = \begin{cases} 1 & n = 1,3,5,\cdots, \\ 0 & n = 2,4,6,\cdots, \end{cases}$$

2. the function χ_2 defined by

$$\begin{aligned} \chi_2(n) &= \cos(\frac{n\pi}{2}) \\ &= \begin{cases} 1 & n = 4l+1, \\ 0 & n = 4l+2, \\ -1 & n = 4l+3, \\ 0 & n = 4l+4, \end{cases} \qquad (l = 0,1,2,3,\cdots.) \end{aligned}$$

3. the function χ_3 defined by

$$\chi_3(n) = \begin{cases} \frac{\sin(n\pi/3)}{\sin(\pi/3)} & n = 1,3,5,\cdots \\ 0 & n = 2,4,6,\cdots \end{cases}$$

$$= \begin{cases} 1 & n = 6l+1, \\ 0 & n = 6l+2, \\ 0 & n = 6l+3, \\ 0 & n = 6l+4, \\ -1 & n = 6l+5, \\ 0 & n = 6l+6, \end{cases} \qquad (l = 0, 1, 2, 3, \cdots.)$$

4. the function χ_4 defined by

$$\chi_4(n) = \begin{cases} \frac{\sin(n\pi/4)}{\sin(\pi/4)} & n = 1, 3, 5, \cdots \\ 0 & n = 2, 4, 6, \cdots \end{cases}$$

$$= \begin{cases} 1 & n = 8l+1, \\ 0 & n = 8l+2, \\ 1 & n = 8l+3, \\ 0 & n = 8l+4, \\ -1 & n = 8l+5, \\ 0 & n = 8l+6, \\ -1 & n = 8l+7, \\ 0 & n = 8l+8. \end{cases} \qquad (l = 0, 1, 2, 3, \cdots.)$$

Proposition 1.1 If both f_1 and f_2 are completely multiplicative, so is the function f defined by

$$f(n) = f_1(n) f_2(n). \tag{1.12}$$

Proof. This proposition follows from the definition plainly. □

In this book, the symbol □ denotes the end of a proof.

Example. From this proposition, we obtain that the functions

$$B_1(n) = \frac{\chi_1(n)}{n},$$
$$B_2(n) = \frac{\chi_2(n)}{n},$$
$$B_3(n) = \frac{\chi_3(n)}{n^2},$$
$$B_4(n) = \frac{\chi_4(n)}{n^2},$$

are completely multiplicative.

1.3 Dirichlet Multiplication

Definition. For two given arithmetical functions f and g, their Dirichlet product is the arithmetical function h defined by the equation [2]

$$h(n) = \sum_{d|n} f(d) g(\frac{n}{d}). \tag{1.13}$$

We write simply

$$h = f * g. \tag{1.14}$$

Example. In the notation of Dirichlet multiplication, the relation (1.7) can be written as

$$\mu * \epsilon = I. \tag{1.15}$$

Proposition 1.2 We have

$$d^2(n) \leq \epsilon^{4*}(n), \tag{1.16}$$

$$d^3(n) \leq \epsilon^{8*}(n), \tag{1.17}$$

for $n = 1, 2, 3, \cdots$.

Here

$$\epsilon^{4*} = \epsilon * \epsilon * \epsilon * \epsilon = d * d,$$

$$\epsilon^{8*} = d * d * d * d.$$

Proof.

From the property of the divisor function $d(m)d(n) \geq d(mn)$, it follows that

$$d * d\ (n) = \sum_{m|n} d(m) d\left(\frac{n}{m}\right) \geq \sum_{m|n} d(n) = d^2(n),$$

Furthermore, we have

$$d * d * d * d\ (n) \geq d^2 * d^2\ (n) \geq d^3(n). \qquad \square$$

Definition. For a given arithmetical function f, if there exists a function f^{-1} such that

$$f * f^{-1} = I, \tag{1.18}$$

then f^{-1} is called the Dirichlet inverse of f.
It can be proved that a function f with $f(1) \neq 0$ has an unique inversion [2].
Example. The Möbius function $\mu(n)$ and the function $\epsilon(n)$ are inverse mutually.

Proposition 1.3 Let f, g be arithmetical functions, and h be their Dirichlet product, i.e.,

$$h = f * g. \tag{1.19}$$

Then we have

$$h = 0 \Longleftrightarrow f = 0, \text{ or } g = 0, \tag{1.20}$$

where 0 is the zero function defined by (1.2).

Proof. From the definition of Dirichlet multiplication, it follows obviously that

$$f = 0, \text{ or } g = 0 \Longrightarrow h = 0. \tag{1.21}$$

Now let us consider

$$h = 0 \Longrightarrow f = 0, \text{ or } g = 0. \tag{1.22}$$

There are three cases.
1. If $f(1) \neq 0$, then $g = h * f^{-1} = 0$. Similarly, if $g(1) \neq 0$, then $f = h * g^{-1} = 0$.
Therefore the relation (1.22) holds in this case.
2. Suppose that

$$f(n) = g(n) = 0 \quad \text{for} \quad n < k, \tag{1.23}$$

and that $f(k) \neq 0$, where k is a positive integer greater than 1.
Then we have

$$\begin{aligned} h(k^2) = f(k)g(k) = 0 \quad &\Longrightarrow \quad g(k) = 0, \\ h(k(k+1)) = f(k)g(k+1) = 0 \quad &\Longrightarrow \quad g(k+1) = 0, \end{aligned}$$

$$\cdots \quad \cdots \quad \cdots$$

$$h(k(k+i)) = f(k)g(k+i) = 0 \implies g(k+i) = 0,$$

$$\cdots \quad \cdots \quad \cdots$$

where i is any nonnegative integer. Therefore we have $g = 0$.

For the same reason, if

$$f(n) = g(n) = 0 \quad \text{for} \quad n < k \tag{1.24}$$

and $g(k) \neq 0$, we have $f = 0$.

Therefore the relation (1.22) holds in this case as well.

3. Suppose that

$$f(n) = g(n) = 0 \quad \text{for any positive integer } n. \tag{1.25}$$

In fact we already have $f = g = 0$ and therefore the relation (1.22) holds trivially in this case.

In one word, the relation (1.22) holds always.

This completes the proof of this proposition. □

It is well-known that if f is completely multiplicative, then

$$f^{-1}(n) = \mu(n)f(n). \tag{1.26}$$

Generally, it is somewhat difficult to calculate the inverse of a function f when f is not completely multiplicative. The following is an example.

Proposition 1.4 The inverse function of θ is given by

$$\theta^{-1}(n) = \begin{cases} \mu(n) & l = 0, \\ 2^{l-1}\mu(2k+1) & l = 1, 2, 3, \cdots, \end{cases} \tag{1.27}$$

where l and k are determined by

$$n = 2^l(2k+1). \tag{1.28}$$

Proof.

Let us introduce another function h, which is defined by

$$h(n) = \begin{cases} n & n = 2^m,\ m = 0, 1, 2, \cdots, \\ 0 & \text{otherwise.} \end{cases} \tag{1.29}$$

Obviously, h is completely multiplicative, and we have

$$h^{-1}(n) = \mu(n)h(n) = \begin{cases} 1 & n = 1, \\ -2 & n = 2, \\ 0 & n = 3, 4, 5, \cdots. \end{cases} \tag{1.30}$$

It is easy to obtain that

$$\theta = \epsilon * h^{-1}. \tag{1.31}$$

Therefore we have

$$\theta^{-1} = \mu * h. \tag{1.32}$$

Furthermore the function θ^{-1} is given explicitly by

$$\theta^{-1}(n) = \begin{cases} \mu(n) & l = 0, \\ 2^{l-1}\mu(2k+1) & l = 1, 2, 3, \cdots, \end{cases} \tag{1.33}$$

where l and k are determined by

$$n = 2^l(2k+1). \quad \square \tag{1.34}$$

1.4 Dirichlet Series

Definition. If f(n) is an arithmetical function, then the series

$$F(s) = \sum_{n=1}^{\infty} \frac{f(n)}{n^s} \tag{1.35}$$

is called the Dirichlet series associated with $f(n)$ [1]. Here s is a positive real number.

If $f(n)$ is multiplicative, then

$$F(s) = \prod_p \left(1 + \frac{f(p)}{p^s} + \frac{f(p^2)}{p^{2s}} + \cdots\right), \tag{1.36}$$

where p runs over all the primes.
Also if $f(n)$ is completely multiplicative, then

$$F(s) = \prod_p \left(1 - \frac{f(p)}{p^s}\right)^{-1}. \tag{1.37}$$

Example. The Dirichlet series associated with the function ϵ is called Riemann zeta function, i.e.,

$$\zeta(s) = \sum_{n=1}^{\infty} \frac{1}{n^s}. \tag{1.38}$$

We have $\zeta(2) = \frac{\pi^2}{6}$, and $\zeta(4) = \frac{\pi^4}{90}$.
Example. The Dirichlet series associated with $\mu(n)$, $d(n)$, $\mid \mu(n) \mid$ are

$$\sum_{n=1}^{\infty} \frac{\mu(n)}{n^s} = \frac{1}{\zeta(s)}, \tag{1.39}$$

$$\sum_{n=1}^{\infty} \frac{d(n)}{n^s} = \zeta^2(s), \tag{1.40}$$

$$\sum_{n=1}^{\infty} \frac{\mid \mu(n) \mid}{n^s} = \frac{\zeta(s)}{\zeta(2s)}. \tag{1.41}$$

Proposition 1.5

$$\frac{1}{1^2} + \frac{d^2(3)}{3^2} + \frac{d^2(5)}{5^2} + \cdots = \frac{3}{128}\pi^4. \tag{1.42}$$

Proof. In fact we have

$$\begin{aligned} &\frac{1}{1^2} + \frac{d^2(2)}{2^2} + \frac{d^2(3)}{3^2} + \cdots + \frac{d^2(n)}{n^2} + \cdots \\ &= \prod_p \left(1 + \frac{d^2(p)}{p^2} + \frac{d^2(p^2)}{p^4} + \frac{d^2(p^m)}{p^{2m}} + \cdots\right) \\ &= \prod_p \left(1 + \frac{2^2}{p^2} + \frac{3^2}{p^4} + \frac{(m+1)^2}{p^{2m}} + \cdots\right) \end{aligned}$$

$$= \prod_p \frac{1+\frac{1}{p^2}}{(1-\frac{1}{p^2})^3} = \prod_p \frac{1-\frac{1}{p^4}}{(1-\frac{1}{p^2})^4}$$
$$= \frac{\zeta^4(2)}{\zeta(4)} = \frac{5}{72}\pi^4.$$

Since

$$\frac{1-\frac{1}{2^4}}{(1-\frac{1}{2^2})^4} = \frac{80}{27},$$

we have

$$\frac{1}{1^2} + \frac{d^2(3)}{3^2} + \frac{d^2(5)}{5^2} + \cdots$$
$$= \frac{5}{72}\pi^4 \frac{27}{80} = \frac{3}{128}\pi^4. \quad \square$$

Proposition 1.6

$$\sum_{n=1}^{\infty} \frac{\mid \theta^{-1}(n) \mid}{n^2} = \frac{18}{\pi^2}. \tag{1.43}$$

Proof.

From (1.32), it follows that θ^{-1} is multiplicative. Therefore we have

$$\sum_{n=1}^{\infty} \frac{\mid \theta^{-1}(n) \mid}{n^2}$$
$$= \prod_p \left(1 + \frac{\mid \theta^{-1}(p) \mid}{p^2} + \frac{\mid \theta^{-1}(p^2) \mid}{p^4} + \frac{\mid \theta^{-1}(p^3) \mid}{p^6} + \cdots\right)$$
$$= (1 + \frac{1}{2^2} + \frac{2}{2^4} + \frac{4}{2^6} + \cdots)(1 + \frac{1}{3^2})(1 + \frac{1}{5^2})(1 + \frac{1}{7^2})(1 + \frac{1}{11^2}) \cdots$$
$$= \frac{3}{2} \frac{1}{(1+\frac{1}{2^2})} \sum_{n=1}^{\infty} \frac{\mid \mu(n) \mid}{n^2}$$
$$= \frac{6}{5} \frac{\frac{\pi^2}{6}}{\frac{\pi^4}{90}} = \frac{18}{\pi^2}. \quad \square$$

The concepts and notations which appear here will be used frequently in the following chapters. When necessary, we shall use the results of this chapter directly without proofs.

Now we are turning to square wave analysis.

Chapter 2

SQUARE WAVE ANALYSIS

About two hundred years ago, the French mathematician J. B. J. Fourier asserted that any function with period 2π can be expressed as a trigonometric series. This great idea has had an important influence upon many sciences including mathematics and physics. In electronics, Fourier analysis has been playing an important role, and a signal is often considered as a superposition of many sine and cosine waves with different frequencies. However, with the development of switching circuits and pulse technique, it becomes very easy to generate and process square waves. For example, it's quite convenient to obtain a system of square waves with different frequencies from a common high-frequency pulse by means of counters, whereas it's complicated to do the same thing with sine and cosine functions. Again, multipliers and summers for square waves are very simple. The following basic questions arise naturally:

1. Can a signal be considered as a superposition of square waves with different frequencies (or periods)?
2. How to decompose a signal into a square wave series?

3. How to approximate a signal best by a superposition of finite square waves?
4. Can this idea bring us new technique?

This problem is called square wave analysis, which is a natural generalization of Fourier's idea.

In this chapter we shall develop a new theory on square waves with different frequencies in $L^2[-\pi,\pi]$ and give an example of applications. To answer the above questions we shall consider the following items:

1. Square wave system and its basic properties,
2. Biorthogonal functions and square wave series,
3. Orthogonalization and the best approximation,
4. An example of applications.

In this book $L^2[-\pi,\pi]$ denotes the Hilbert space of real quadratically integrable functions defined on $[-\pi,\pi]$, and $\langle,\rangle$ the inner product in the space, $||\bullet||$ the norm in the space.

2.1 Square Wave System and its Basic Properties

Definition. The functions $X(x)$ and $Y(x)$ are called an even and odd square wave, respectively, which are defined on $(-\infty,+\infty)$ by

$$X(x) = \frac{\pi}{4}\operatorname{sgn}(\cos x), \tag{2.1}$$

and

$$Y(x) = \frac{\pi}{4}\operatorname{sgn}(\sin x). \tag{2.2}$$

Here sgn is the sign function:

$$\operatorname{sgn}(x) = \begin{cases} 1 & x > 0, \\ 0 & x = 0, \\ -1 & x < 0. \end{cases} \tag{2.3}$$

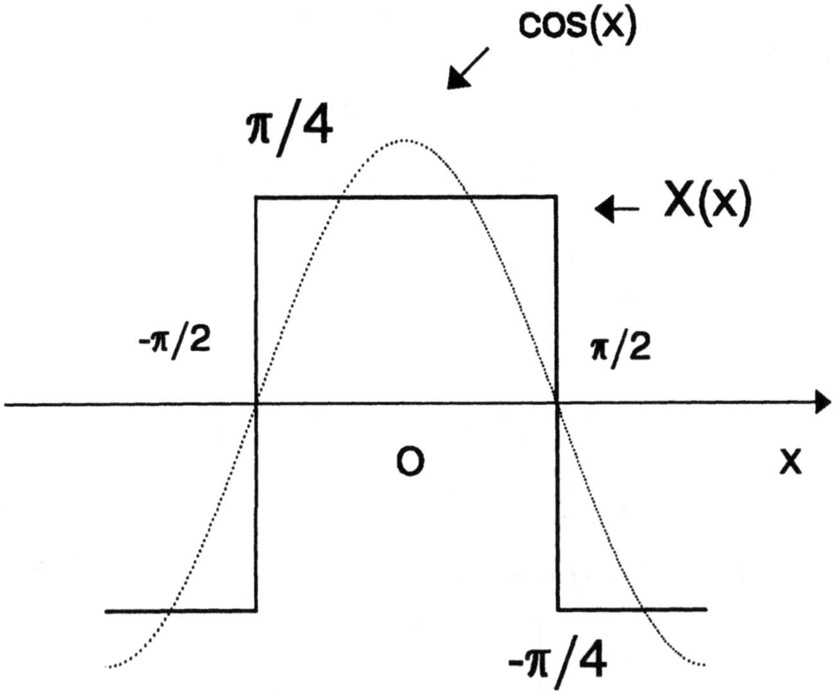

Figure 2.1: The even square wave $X(x)$.

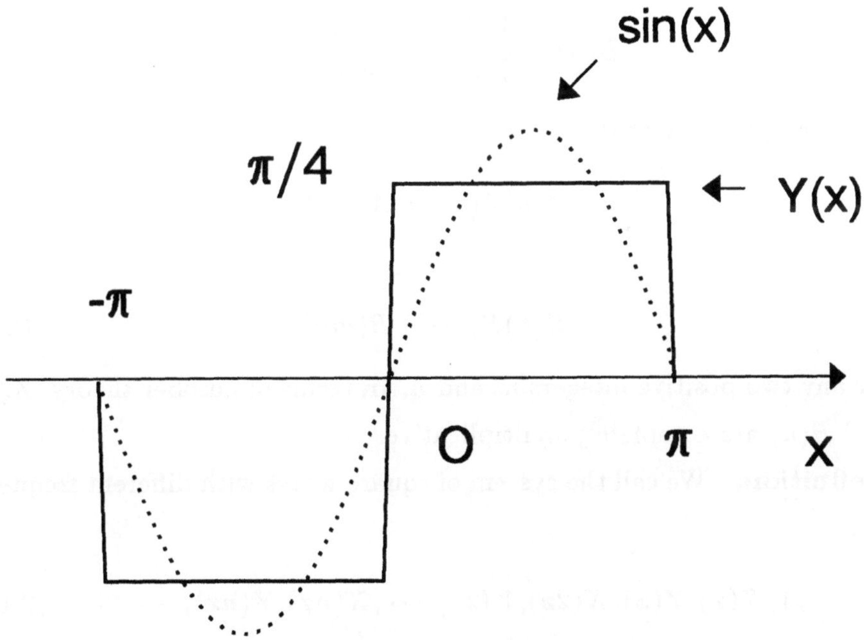

Figure 2.2: The odd square wave $Y(x)$.

For convenience let the amplitude of the square waves be $\frac{\pi}{4}$. See Figure 2.1 and Figure 2.2

The Fourier series of the square waves are [3]

$$\begin{aligned} X(x) &= \cos x - \frac{1}{3}\cos(3x) + \frac{1}{5}\cos(5x) - \frac{1}{7}\cos(7x) + \cdots \\ &= \sum_{n=1}^{\infty} A(n)\cos(nx), \end{aligned} \tag{2.4}$$

and

$$\begin{aligned} Y(x) &= \sin x + \frac{1}{3}\sin(3x) + \frac{1}{5}\sin(5x) + \frac{1}{7}\sin(7x) + \cdots \\ &= \sum_{n=1}^{\infty} B(n)\sin(nx). \end{aligned} \tag{2.5}$$

Notice that the Fourier coefficients

$$A(n) = \begin{cases} (-1)^{\frac{n-1}{2}}\,\frac{1}{n} & n = 1,3,5,\cdots \\ 0 & n = 2,4,6,\cdots \end{cases} \tag{2.6}$$

and

$$B(n) = \begin{cases} \frac{1}{n} & n = 1,3,5,\cdots \\ 0 & n = 2,4,6,\cdots \end{cases} \tag{2.7}$$

have the property that

$$A(n)A(m) = A(mn), \tag{2.8}$$

and

$$B(n)B(m) = B(mn), \tag{2.9}$$

for any two positive integers m and n. In terms of number theory, $A(n)$ and $B(n)$ are completely multiplicative.

Definition. We call the system of square waves with different frequencies

$$1, X(x), Y(x), X(2x), Y(2x), \cdots, X(nx), Y(nx), \cdots \tag{2.10}$$

the square wave system.

By the way, please note that the square wave system defined here is quite different from the Walsh function system, which consists of very complicated rectangular waves [4][5]. The Redemacher function system can be considered as a proper subset of the square wave system in essence, which is a well-known incomplete system [5].

In what follows we shall show that the square wave system is linearly independent, nonorthogonal and complete in $L^2[-\pi, \pi]$.

(1) Linear independence

Proposition 2.1 The function system of odd square waves

$$Y(x), Y(2x), Y(3x), \cdots, Y(nx), \cdots \tag{2.11}$$

is linearly independent, i.e., any finite subsystem of the function system is linearly independent.

Proof. We'll prove that $\{Y(nx)\}_{n=1}^{N}$ is linearly independent for $N = 2, 3, 4, \cdots$ by mathematical induction.

1^0. $Y(x)$ and $Y(2x)$ are linearly independent.

Notice that

$$c_1 Y(x) + c_2 Y(2x) = \begin{cases} \frac{\pi}{4}(c_1 + c_2) & x \in (0, \frac{\pi}{2}), \\ \frac{\pi}{4}(c_1 - c_2) & x \in (\frac{\pi}{2}, \pi). \end{cases} \tag{2.12}$$

Therefore,

$$c_1 Y(x) + c_2 Y(2x) = 0$$

$$\Rightarrow \begin{cases} (c_1 + c_2) = 0 \\ (c_1 - c_2) = 0 \end{cases}$$

$$\Rightarrow \quad c_1 = c_2 = 0.$$

2^0. If $\{Y(nx)\}_{n=1}^{k}$ is linearly independent, then so is $\{Y(nx)\}_{n=1}^{k+1}$.

Notice that

$$\begin{aligned} & c_1 Y(x) + c_2 Y(2x) + \cdots + c_k Y(kx) + c_{k+1} Y((k+1)x) \\ & = \begin{cases} \frac{\pi}{4}(c_1 + c_2 + \cdots + c_k + c_{k+1}) & x \in (0, \frac{\pi}{k+1}), \\ \frac{\pi}{4}(c_1 + c_2 + \cdots + c_k - c_{k+1}) & x \in (\frac{\pi}{k+1}, \frac{\pi}{k}). \end{cases} \end{aligned} \tag{2.13}$$

Therefore,

$$c_1Y(x) + c_2Y(2x) + \cdots + c_kY(kx) + c_{k+1}Y((k+1)x) = 0$$

$$\begin{aligned}
&\Rightarrow \begin{cases} c_1 + c_2 + \cdots + c_k + c_{k+1} = 0 \\ c_1 + c_2 + \cdots + c_k - c_{k+1} = 0 \end{cases} \\
&\Rightarrow c_{k+1} = 0 \\
&\Rightarrow c_1Y(x) + c_2Y(2x) + \cdots + c_kY(kx) = 0 \\
&\Rightarrow c_1 = c_2 = \cdots = c_k = 0 \\
&\Rightarrow c_1 = c_2 = \cdots = c_k = c_{k+1} = 0.
\end{aligned}$$

From 1^0 and 2^0, it follows that the function system of odd square waves $\{Y(nx)\}_{n=1}^{\infty}$ is linearly independent. □

It can be proved in the same way that the function system of even square waves

$$1, X(x), X(2x), \cdots, X(nx), \cdots \tag{2.14}$$

is linearly independent as well. Considering the orthogonality of even functions and odd functions, we have

Proposition 2.2 The square wave system is linearly independent.

(2) Nonorthogonality

Though odd square waves and even square waves are orthogonal, the square wave system is not an orthogonal system. In fact, we have the following inner product formulas.

Proposition 2.3 For any two positive integers m and n, we have

$$\langle X(mx), X(nx)\rangle = \frac{\pi^3}{8}A\left(\frac{mn}{(m,n)^2}\right), \tag{2.15}$$

$$\langle Y(mx), Y(nx)\rangle = \frac{\pi^3}{8}B\left(\frac{mn}{(m,n)^2}\right), \tag{2.16}$$

where (m, n) denotes the greatest common divisor of m and n.

Proof.

$$\begin{aligned}
&\langle Y(mx), Y(nx)\rangle \\
&= \langle \sum_{l=1}^{\infty} B(l)\sin(mlx), \sum_{k=1}^{\infty} B(k)\sin(nkx)\rangle \\
&= \pi \sum_{ml=nk} B(l)B(k) \\
&= \pi \sum_{r=1}^{\infty} B(\tfrac{r[m,n]}{m})B(\tfrac{r[m,n]}{n}) \\
&= \pi B(\tfrac{[m,n]^2}{mn}) \sum_{r=1}^{\infty} B(r^2) \\
&= \tfrac{\pi^3}{8} B(\tfrac{mn}{(m,n)^2}).
\end{aligned}$$

Here $[m, n]$ denotes the least common multiple of m and n, and we have

$$mn = (m,n)[m,n]. \tag{2.17}$$

The other equality can be deduced similarly. □

(3) Completeness

Proposition 2.4 The two series

$$\sin x = \sum_{n=1}^{\infty} \mu(n)B(n)Y(nx) \tag{2.18}$$

and

$$\cos x = \sum_{n=1}^{\infty} \mu(n)A(n)X(nx) \tag{2.19}$$

converge unconditionally in $L^2[-\pi, \pi]$.

Proof.

$$\sum_{n=1}^{\infty} \mu(n)B(n)Y(nx)$$

$$= \sum_{n=1}^{\infty} \mu(n)B(n) \sum_{m=1}^{\infty} B(m)\sin(mnx) \tag{2.20}$$

$$= \sum_{k=1}^{\infty} (\sum_{d|k} \mu(d))B(k)\sin kx \tag{2.21}$$

$$= \sin x.$$

From (2.20) to (2.21) the summation order was changed, but this doesn't matter for the series encountered here are convergent unconditionally in

$L^2[-\pi,\pi]$. The reason for unconditional convergence is that the series

$$\sum_{k=1}^{\infty}\left(\sum_{d|k} | \mu(d) | \right) | B(k) | \ \sin kx \tag{2.22}$$

is convergent, the norm square of which is

$$\begin{aligned} &\pi \sum_{k=1}^{\infty}\left(\sum_{d|k} | \mu(d) | \right)^2 B^2(k) \\ &\leq \pi \sum_{k=1}^{\infty} d^2(k)B^2(k) \\ &= \pi[\tfrac{1}{1^2} + \tfrac{d^2(3)}{3^2} + \tfrac{d^2(5)}{5^2} + \cdots] \\ &= \tfrac{3}{128}\pi^5. \end{aligned}$$

Here $d(k)$ denotes the number of positive factors of the positive integer k and we have

$$\sum_{d|k} | \mu(d) | \leq d(k). \tag{2.23}$$

The series about $\cos x$ can be discussed similarly. □

The convergence process of the series (2.5) and (2.18) is shown in Figure 2.3. The first few square wave components contained in $\sin x$ are shown in Figure 2.4.

Proposition 2.5 The square wave system is complete in $L^2[-\pi,\pi]$.

Proof. Since the sine and cosine function system is complete in $L^2[-\pi,\pi]$, and sine and cosine functions can be expressed as a square wave series, the square wave system is complete in the Hilbert space $L^2[-\pi,\pi]$. In other words, a superposition of square waves can approximates an arbitrary quadratically integrable function with a vanishing mean-square error. □

Remark. Since the square wave system is complete, one can replace a signal by a linear combination of square waves at an arbitrary accuracy. From the practical viewpoint, one can say that a signal can be considered as a superposition of square waves. However, due to its nonorthogonality, one needs a special method when decomposing a signal into square waves.

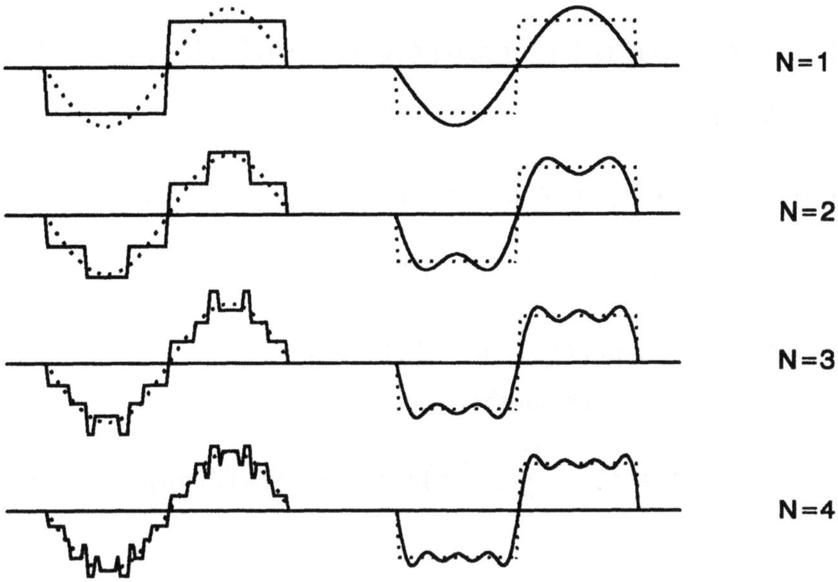

Figure 2.3: The superposition of sine functions $\sum_{k=1}^{N} \frac{1}{2k-1} \sin((2k-1)x)$ approximates the square wave $Y(x)$, and the superposition of square waves $\sum_{k=1}^{N} \frac{\mu(2k-1)}{2k-1} Y((2k-1)x)$ approximates $\sin x$ as their term number N increase.

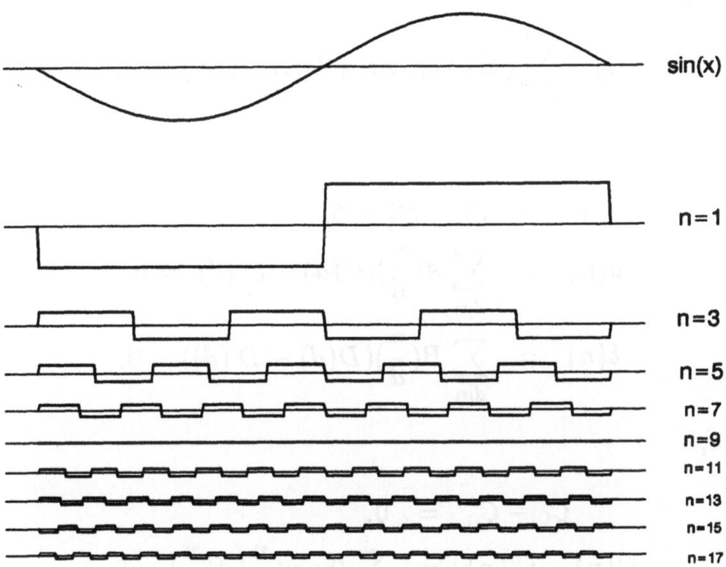

Figure 2.4: The first few square wave components $\mu(n)B(n)Y(nx)$ contained in $\sin x$. ($n = 1, 3, 5, 7, 9, 11, 13, 15, 17$.)

2.2 Biorthogonal Functions and Square Wave Series

Proposition 2.6 If $f(x) \in L^2[-\pi, \pi]$ has a square wave series:

$$f(x) = C_0 + \sum_{n=1}^{\infty}(C(n)X(nx) + D(n)Y(nx)), \tag{2.24}$$

then all the coefficients $C_0, C(n), D(n)$ are unique.

Proof. Suppose $f(x)$ has another square wave series:

$$f(x) = C'_0 + \sum_{n=1}^{\infty}(C'(n)X(nx) + D'(n)Y(nx)). \tag{2.25}$$

So,

$$(C_0 - C'_0) + \sum_{n=1}^{\infty} ((C(n) - C'(n))X(nx) + (D(n) - D'(n))Y(nx)) = 0.$$

Substituting their Fourier series for square waves $X(nx)$ and $Y(nx)$, one obtains that

$$a_0 + \sum_{n=1}^{\infty}(a(n) \cos nx + b(n) \sin nx) = 0, \tag{2.26}$$

where

$$\begin{aligned} a_0 &= C_0 - C'_0 = 0, \\ a(n) &= \sum_{d|n} A(\frac{n}{d})(C(d) - C'(d)) = 0, \\ b(n) &= \sum_{d|n} B(\frac{n}{d})(D(d) - D'(d)) = 0. \end{aligned}$$

Therefore,

$$\begin{aligned} C_0 - C'_0 &= 0, \\ C(n) - C'(n) &= \sum_{d|n} 0 \bullet \mu(\frac{n}{d})A(\frac{n}{d}) = 0, \\ D(n) - D'(n) &= \sum_{d|n} 0 \bullet \mu(\frac{n}{d})B(\frac{n}{d}) = 0. \end{aligned}$$

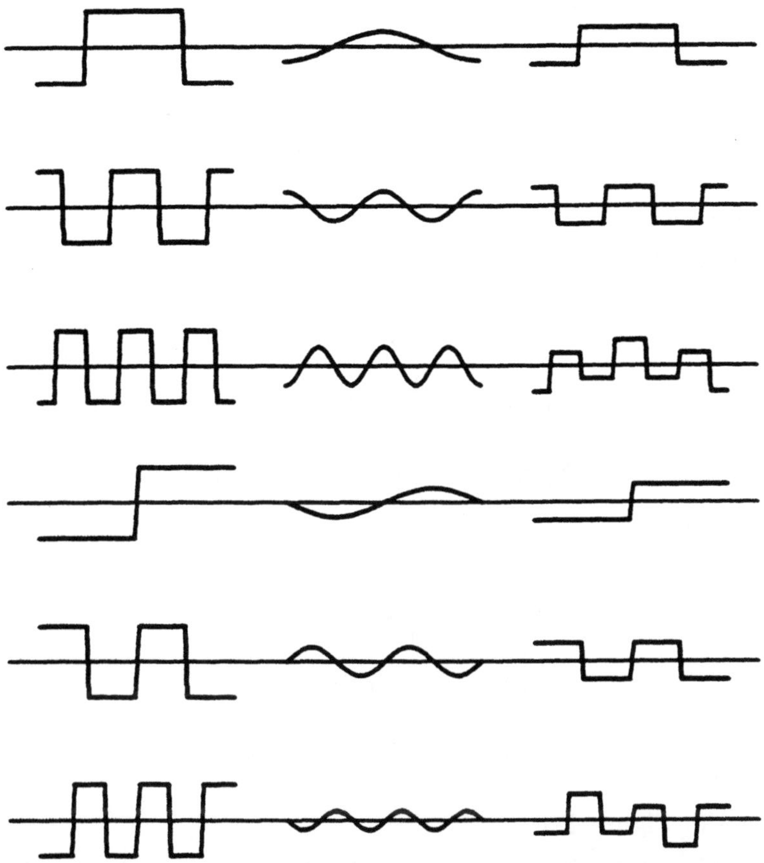

Figure 2.5: Square waves, biorthogonal functions and orthonormalized functions.

$$\begin{pmatrix} X(x) & h_1(x) & E_1(x) \\ X(2x) & h_2(x) & E_2(x) \\ X(3x) & h_3(x) & E_3(x) \\ Y(x) & g_1(x) & e_1(x) \\ Y(2x) & g_2(x) & e_2(x) \\ Y(3x) & g_3(x) & e_3(x) \end{pmatrix}$$

In other words, $C_0, C(n), D(n)$ are unique. □

Definition. For $n = 1, 2, 3, \cdots$, the functions $h_n(x)$ and $g_n(x)$ are defined by

$$h_n(x) = \frac{1}{\pi}\sum_{d|n}\mu(\frac{n}{d})A(\frac{n}{d})\cos dx, \tag{2.27}$$

and

$$g_n(x) = \frac{1}{\pi}\sum_{d|n}\mu(\frac{n}{d})B(\frac{n}{d})\sin dx, \tag{2.28}$$

respectively.

Example. When $n = 1, 2, 3$, we have

$$\begin{aligned} h_1(x) &= \frac{1}{\pi}\cos x, \\ h_2(x) &= \frac{1}{\pi}\cos 2x, \\ h_3(x) &= \frac{1}{\pi}(\cos 3x + \frac{1}{3}\cos x), \end{aligned}$$

and

$$\begin{aligned} g_1(x) &= \frac{1}{\pi}\sin x, \\ g_2(x) &= \frac{1}{\pi}\sin 2x, \\ g_3(x) &= \frac{1}{\pi}(\sin 3x - \frac{1}{3}\sin x), \end{aligned}$$

as shown in Figure 2.5.

Proposition 2.7 For any two positive integers m and n, we have

$$\langle X(mx), h_n(x)\rangle = \delta_{mn}, \tag{2.29}$$

$$\langle Y(mx), g_n(x)\rangle = \delta_{mn}, \tag{2.30}$$

where δ_{mn} is the Kronecker's delta symbol.

Proof.

$$\begin{aligned} &\langle Y(mx), g_n(x)\rangle \\ = &\langle \sum_{k=1}^{\infty} B(k)\sin(mkx), \frac{1}{\pi}\sum_{d|n} B(\frac{n}{d})\mu(\frac{n}{d})\sin dx\rangle \end{aligned}$$

$$
\begin{aligned}
&= \sum_{m|d,d|n} B(\frac{n}{d})\mu(\frac{n}{d})B(\frac{d}{m}) \\
&= \begin{cases} 0 & m \nmid n \\ B(\frac{n}{m}) \sum_{m|d,d|n} \mu(\frac{n}{d}) = \begin{cases} 0 & \frac{n}{m} = 2,3,\cdots \\ 1 & \frac{n}{m} = 1 \end{cases} \end{cases} \\
&= \begin{cases} 0 & m \neq n \\ 1 & m = n \end{cases} \\
&= \delta_{mn}.
\end{aligned}
$$

The other relation in this proposition can be proved similarly. □

Proposition 2.8 The square wave system

$$1, X(x), Y(x), X(2x), Y(2x), \cdots, X(nx), Y(nx), \cdots \tag{2.31}$$

and the function system

$$\frac{1}{2\pi}, h_1(x), g_1(x), h_2(x), g_2(x), \cdots, h_n(x), g_n(x), \cdots \tag{2.32}$$

are biorthogonal.

Proof. Due to Proposition 2.7 and the orthogonality of even functions and odd functions, this proposition is plain. □

Proposition 2.9 If a function $f(x) \in L^2[-\pi, \pi]$ has a square wave series:

$$f(x) = C_0 + \sum_{n=1}^{\infty}(C(n)X(nx) + D(n)Y(nx)), \tag{2.33}$$

then the coefficients in the square wave series can be determined by

$$
\begin{aligned}
C_0 &= \frac{1}{2\pi}\int_{-\pi}^{+\pi} f(x)dx, \\
C(n) &= \int_{-\pi}^{+\pi} f(x)h_n(x)dx, \\
D(n) &= \int_{-\pi}^{+\pi} f(x)g_n(x)dx.
\end{aligned}
$$

Proof. By Proposition 2.8, this proposition is plain. □

Consequence of Proposition 2.9. If a function $f(x) \in L^2[-\pi, \pi]$ has a square wave series:

$$f(x) = a_0 + \sum_{n=1}^{\infty}(a(n)\cos nx + b(n)\sin nx), \tag{2.34}$$

$$= C_0 + \sum_{n=1}^{\infty}(C(n)X(nx) + D(n)Y(nx)), \tag{2.35}$$

then the square wave coefficients and Fourier coefficients are related by

$$\begin{aligned} C_0 &= a_0, \\ C(n) &= \sum_{d|n}\mu(\tfrac{n}{d})A(\tfrac{n}{d})a(d), \\ D(n) &= \sum_{d|n}\mu(\tfrac{n}{d})B(\tfrac{n}{d})b(d), \end{aligned} \tag{2.36}$$

or inversely

$$\begin{aligned} a_0 &= C_0, \\ a(n) &= \sum_{d|n}A(\tfrac{n}{d})C(d), \\ b(n) &= \sum_{d|n}B(\tfrac{n}{d})D(d). \end{aligned} \tag{2.37}$$

In the notation of Dirichlet multiplication, we write

$$C_0 = a_0, \ C = A^{-1} * a, \ D = B^{-1} * b, \tag{2.38}$$

or

$$a_0 = C_0, \ a = A * C, \ b = B * D. \tag{2.39}$$

Now a natural question is: satisfying what conditions a function can be expanded as a square wave series. The following is a sufficient condition.

Proposition 2.10 Let the Fourier series of $f(x) \in L^2[-\pi, \pi]$ be

$$f(x) = a_0 + \sum_{n=1}^{\infty}(a(n)\cos nx + b(n)\sin nx). \tag{2.40}$$

If the Fourier coefficients satisfy the condition

$$\sum_{n=1}^{\infty} a^2(n)d(n) < \infty, \tag{2.41}$$

$$\sum_{n=1}^{\infty} b^2(n)d(n) < \infty, \tag{2.42}$$

the series

$$f(x) = C_0 + \sum_{n=1}^{\infty}(C(n)X(nx) + D(n)Y(nx)), \tag{2.43}$$

converges unconditionally in $L^2[-\pi, \pi]$, where

$$\begin{aligned} C_0 &= a_0, \\ C(n) &= \sum_{d|n} \mu(\tfrac{n}{d})A(\tfrac{n}{d})a(d), \\ D(n) &= \sum_{d|n} \mu(\tfrac{n}{d})B(\tfrac{n}{d})b(d). \end{aligned} \tag{2.44}$$

Proof.

$$\begin{aligned} f(x) &= a_0 + \sum_{n=1}^{\infty}(a(n)\cos nx + b(n)\sin nx) \\ &= a_0 + \sum_{n=1}^{\infty} a(n) \sum_{l=1}^{\infty} A(l)\mu(l)X(nlx) \\ &\quad + \sum_{n=1}^{\infty} b(n) \sum_{l=1}^{\infty} B(l)\mu(l)Y(nlx) \qquad (2.45) \\ &= a_0 + \sum_{n=1}^{\infty}\left(\sum_{d|n} a(d)\mu(\frac{n}{d})A(\frac{n}{d})\right) X(nx) \\ &\quad + \sum_{n=1}^{\infty}\left(\sum_{d|n} b(d)\mu(\frac{n}{d})B(\frac{n}{d})\right) Y(nx) \qquad (2.46) \\ &= C_0 + \sum_{n=1}^{\infty}(C(n)X(nx) + D(n)Y(nx)). \end{aligned}$$

From (2.45) to (2.46), the summation orders were changed. But this does not matter because all the following series

$$\begin{aligned} &\sum_{n=1}^{\infty} a(n) \sum_{l=1}^{\infty} A(l)\mu(l)X(nlx) \\ = &\sum_{n=1}^{\infty} a(n) \sum_{l=1}^{\infty} A(l)\mu(l) \sum_{m=1}^{\infty} A(m)\cos(nlmx) \\ = &\sum_{k=1}^{\infty}\left(\sum_{nlm=k} a(n)A(l)\mu(l)A(m)\right)\cos(kx) \end{aligned}$$

and

$$\begin{aligned}
&\sum_{n=1}^{\infty} b(n) \sum_{l=1}^{\infty} B(l)\mu(l)Y(nlx) \\
=& \sum_{n=1}^{\infty} b(n) \sum_{l=1}^{\infty} B(l)\mu(l) \sum_{m=1}^{\infty} B(m)\sin(nlmx) \\
=& \sum_{k=1}^{\infty} \left(\sum_{nlm=k} b(n)B(l)\mu(l)B(m) \right) \sin(kx)
\end{aligned}$$

are unconditionally convergent in $L^2[-\pi, \pi]$, or one can say, they have sums independent of orders. The reason for the unconditional convergence is that the series

$$\sum_{k=1}^{\infty} \left(\sum_{nlm=k} | a(n)A(l)\mu(l)A(m) | \right) \cos(kx) \tag{2.47}$$

and

$$\sum_{k=1}^{\infty} \left(\sum_{nlm=k} | b(n)B(l)\mu(l)B(m) | \right) \sin(kx) \tag{2.48}$$

are convergent in $L^2[-\pi, \pi]$. As an example, the norm square of the series (2.47) is

$$\begin{aligned}
&\pi \sum_{k=1}^{\infty} \left(\sum_{nlm=k} | a(n)A(l)\mu(l)A(m) | \right)^2 \\
=& \pi \sum_{k=1}^{\infty} \left(\sum_{nj=k} | a(n) | \, | A(j) | \sum_{lm=j} | \mu(l) | \right)^2 \\
=& \pi \sum_{k=1}^{\infty} \left(\sum_{nj=k} | a(n) | \, A'(j) \right)^2 \\
\leq& \pi \sum_{k=1}^{\infty} d(k) \sum_{nj=k} a^2(n) A'^{\,2}(j) && (2.49) \\
\leq& \pi \sum_{k=1}^{\infty} \sum_{nj=k} a^2(n)d(n) A'^{\,2}(j) d(j) && (2.50) \\
=& \pi \sum_{n=1}^{\infty} a^2(n)d(n) \sum_{j=1}^{\infty} A'^{\,2}(j) d(j) && (2.51) \\
<& \infty, && (2.52)
\end{aligned}$$

where

$$A'(j) = \mid A(j) \mid \sum_{l|j} \mid \mu(l) \mid . \tag{2.53}$$

From (2.49) to (2.50) we used

$$d(nj) \le d(n)d(j), \tag{2.54}$$

and from (2.51) to (2.52) we used

$$\begin{aligned} & \sum_{j=1}^{\infty} A'^{\,2}(j)d(j) \\ = & \sum_{j=1}^{\infty} A^2(j)d(j)(\sum_{d|j} \mid \mu(d) \mid)^2 \\ \le & \sum_{j=1}^{\infty} A^2(j)d^3(j) \\ < & \ \infty. \quad \square \end{aligned}$$

Definition. A subset K of $L^2[-\pi, \pi]$ is defined by

$$K = \{ \ f(x) \in L^2[-\pi,\pi] : \ f(x) = a_0 + \sum_{n=1}^{\infty} (a(n) \cos nx + b(n) \sin nx) \\ \text{with} \sum_{n=1}^{\infty} a^2(n)d(n) < \infty, \ \sum_{n=1}^{\infty} b^2(n)d(n) < \infty \ \}.$$

Proposition 2.11 K is a linear subspace of $L^2[-\pi, \pi]$.

Proof. One can prove the following two items easily.

1^0. If $f(x) \in K$, then $cf(x) \in K$, where c is any real number.

2^0. If $f_1(x), f_2(x) \in K$,then $f_1(x) + f_2(x) \in K$. $\square$

Proposition 2.12 The square wave system is an unconditional basis of the subspace K.

Proof. First of all, the square wave system is in K.

As an example, let us consider

$$Y(mx) = \sum_{l=1}^{\infty} B(l) \sin(lmx) = \sum_{n=1}^{\infty} b(n) \sin(nx). \tag{2.55}$$

It follows that

$$
\begin{aligned}
& \sum_{n=1}^{\infty} b^2(n)d(n) \\
= & \sum_{l=1}^{\infty} B^2(l)d(ml) \\
\leq & \; d(m)\sum_{l=1}^{\infty} B^2(l)d(l) \\
< & \; \infty.
\end{aligned}
$$

Secondly, by Proposition 2.6 and Proposition 2.10, the square wave system is an unconditional basis of the subspace K. □

Whether the square wave system is a basis of $L^2[-\pi,\pi]$ is still unknown.

Remark. Though we do not know whether the square wave system is a basis of $L^2[-\pi,\pi]$, the real signal always has little components with high frequencies, so from the practical viewpoint one can say that any signal can be decomposed into a square wave series by means of the biorthogonal functions.

2.3 Orthogonalization and the Best Approximation

Now that the square wave system is linearly independent, nonorthogonal and complete, one can get an orthonormal basis of $L^2[-\pi,\pi]$ by the Gram-Schmidt orthogonalization process.

Definition. The function $\beta(n)$ is defined on positive integers by

$$\beta(n) = \sum_{d|n} \mu(d)A^2(d) = \sum_{d|n} \mu(d)B^2(d). \tag{2.56}$$

For example, $\beta(1) = 1,\ \beta(2) = 1,\ \beta(3) = 1 - (\frac{1}{3})^2 = \frac{8}{9}$.

Proposition 2.13 For $n = 1, 2, 3, \cdots$, we have

$$\sum_{d|n} \beta(d)A^2(\frac{n}{d}) = \sum_{d|n} \beta(d)B^2(\frac{n}{d}) = 1. \tag{2.57}$$

Proof. In the notation of Dirichlet multiplication, we have

$$\beta = (A^2)^{-1} * \epsilon. \tag{2.58}$$

Therefore it follows that

$$\beta * A^2 = \epsilon. \tag{2.59}$$

This is the conclusion we want. □

Proposition 2.14 $\beta(n) > 0$ for $n = 1, 2, 3, \cdots$.

Proof. From (2.58), it follows that β is multiplicative and that

$$\beta(n) = \prod_{p|n} \left(1 - A^2(p)\right),$$

where p runs over the prime divisors of n. Obviously $\beta(n) > 0$. □

Definition. For $n = 1, 2, 3, \cdots$, the functions $E_n(x)$ and $e_n(x)$ are defined by

$$E_n(x) = \frac{1}{\sqrt{\frac{\pi^3}{8}}} \frac{1}{\sqrt{\beta(n)}} \sum_{d|n} \mu(\frac{n}{d}) A(\frac{n}{d}) X(dx), \tag{2.60}$$

and

$$e_n(x) = \frac{1}{\sqrt{\frac{\pi^3}{8}}} \frac{1}{\sqrt{\beta(n)}} \sum_{d|n} \mu(\frac{n}{d}) B(\frac{n}{d}) Y(dx), \tag{2.61}$$

respectively.

Example. For $n = 1, 2, 3$, we have

$$\begin{aligned} E_1(x) &= \frac{1}{\sqrt{\frac{\pi^3}{8}}} X(x), \\ E_2(x) &= \frac{1}{\sqrt{\frac{\pi^3}{8}}} X(2x), \\ E_3(x) &= \frac{1}{\sqrt{\frac{\pi^3}{9}}} (X(3x) + \frac{1}{3} X(x)), \end{aligned}$$

and

$$e_1(x) = \frac{1}{\sqrt{\frac{\pi^3}{8}}} Y(x),$$

$$
\begin{aligned}
e_2(x) &= \frac{1}{\sqrt{\frac{\pi^3}{8}}} Y(2x), \\
e_3(x) &= \frac{1}{\sqrt{\frac{\pi^3}{9}}} (Y(3x) - \frac{1}{3} Y(x)),
\end{aligned}
$$

as shown in Figure 2.5.

Proposition 2.15 Both $\{E_n(x)\}_{n=1}^{\infty}$ and $\{e_n(x)\}_{n=1}^{\infty}$ are orthonormal systems (ONS).

Proof.

$$
\begin{aligned}
&\langle E_m(x), E_n(x)\rangle \\
&= \frac{1}{\frac{\pi^3}{8}} \frac{1}{\sqrt{\beta(m)\beta(n)}} \sum_{d|m} A(\tfrac{m}{d})\mu(\tfrac{m}{d}) \sum_{d'|n} A(\tfrac{n}{d'})\mu(\tfrac{n}{d'}) \langle X(dx), X(d'x)\rangle \\
&= \frac{1}{\sqrt{\beta(m)\beta(n)}} \sum_{d|m} A(\tfrac{m}{d})\mu(\tfrac{m}{d}) \sum_{d'|n} A(\tfrac{n}{d'})\mu(\tfrac{n}{d'}) A(\tfrac{dd'}{(d,d')^2}) \\
&= \frac{1}{\sqrt{\beta(m)\beta(n)}} \sum_{d|m} A(\tfrac{m}{d})\mu(\tfrac{m}{d}) \sum_{d'|n} A(\tfrac{n}{d'})\mu(\tfrac{n}{d'}) A(\tfrac{dd'}{(d,d')^2}) \sum_{l|(d,d')} \beta(l) A^2(\tfrac{(d,d')}{l}) \\
&= \frac{1}{\sqrt{\beta(m)\beta(n)}} \sum_{d|m} A(\tfrac{m}{d})\mu(\tfrac{m}{d}) \sum_{d'|n} A(\tfrac{n}{d'})\mu(\tfrac{n}{d'}) \sum_{l|(d,d')} \beta(l) A(\tfrac{d}{l}) A(\tfrac{d'}{l}) \\
&= \frac{1}{\sqrt{\beta(m)\beta(n)}} \sum_{d|m} A(\tfrac{m}{d})\mu(\tfrac{m}{d}) \sum_{l|d} \beta(l) A(\tfrac{d}{l}) \sum_{l|d', d'|n} A(\tfrac{n}{d'}) A(\tfrac{d'}{l}) \mu(\tfrac{n}{d'}) \\
&= \frac{1}{\sqrt{\beta(m)\beta(n)}} \sum_{d|m} A(\tfrac{m}{d})\mu(\tfrac{m}{d}) \sum_{l|d} \beta(l) A(\tfrac{d}{l}) \delta_{nl} \\
&= \frac{1}{\sqrt{\beta(m)\beta(n)}} \sum_{n|d, d|m} A(\tfrac{m}{d})\mu(\tfrac{m}{d}) A(\tfrac{d}{n}) \beta(n) \\
&= \delta_{mn}.
\end{aligned}
$$

Similarly, one has

$$
\langle e_m(x), e_n(x)\rangle = \delta_{mn}. \quad \square
$$

Proposition 2.16 The function system

$$
\frac{1}{\sqrt{2\pi}}, E_1(x), e_1(x), E_2(x), e_2(x), \cdots, E_n(x), e_n(x), \cdots \tag{2.62}
$$

is an orthonormal basis of $L^2[-\pi, \pi]$.

Proof. The square wave system is linearly independent and complete, so its orthonormalized function system is an orthonormal basis of $L^2[-\pi, \pi]$.

□

Consequence of Proposition 2.16 Any function $f(x) \in L^2[-\pi, \pi]$ can be written as the following series:

$$f(x) = r_0 + \sum_{n=1}^{\infty}(r(n)E_n(x) + q(n)e_n(x)), \tag{2.63}$$

where

$$r_0 = \frac{1}{2\pi}\int_{-\pi}^{+\pi} f(x)dx, \tag{2.64}$$

$$r(n) = \int_{-\pi}^{+\pi} f(x)E_n(x)dx, \tag{2.65}$$

$$q(n) = \int_{-\pi}^{+\pi} f(x)e_n(x)dx. \tag{2.66}$$

Proposition 2.17

$$\sin x = \frac{\pi}{\sqrt{\frac{\pi^3}{8}}}\sum_{n=1}^{\infty}\frac{1}{\sqrt{\beta(n)}}\mu(n)B(n)e_n(x), \tag{2.67}$$

$$\cos x = \frac{\pi}{\sqrt{\frac{\pi^3}{8}}}\sum_{n=1}^{\infty}\frac{1}{\sqrt{\beta(n)}}\mu(n)A(n)E_n(x). \tag{2.68}$$

Proof.

It is because that

$$\begin{aligned}
&\langle \sin x, e_n(x)\rangle \\
&= \frac{1}{\sqrt{\frac{\pi^3}{8}}}\frac{1}{\sqrt{\beta(n)}}\sum_{d|n}\mu(\frac{n}{d})B(\frac{n}{d})\langle \sin x, Y(dx)\rangle \\
&= \frac{\pi}{\sqrt{\frac{\pi^3}{8}}}\frac{1}{\sqrt{\beta(n)}}\mu(n)B(n)
\end{aligned}$$

and

$$\langle \cos x, E_n(x)\rangle = \frac{\pi}{\sqrt{\frac{\pi^3}{8}}}\frac{1}{\sqrt{\beta(n)}}\mu(n)A(n). \quad \square$$

Next let us consider how to approximate a given function $f(x) \in L^2[-\pi, \pi]$ by a superposition of finite square waves.

Definition. Let N be a given positive integer. For $n = 1, 2, 3, \cdots, N$, one defines

$$\begin{aligned} h_n^{(N)}(x) &= \frac{1}{\frac{\pi^3}{8}} \sum_{m=1}^{N} \left(\sum_{\substack{1 \le k \le N \\ m|k, n|k}} \frac{1}{\beta(k)} A(\frac{k}{m}) A(\frac{k}{n}) \mu(\frac{k}{m}) \mu(\frac{k}{n}) \right) X(mx), \\ g_n^{(N)}(x) &= \frac{1}{\frac{\pi^3}{8}} \sum_{m=1}^{N} \left(\sum_{\substack{1 \le k \le N \\ m|k, n|k}} \frac{1}{\beta(k)} B(\frac{k}{m}) B(\frac{k}{n}) \mu(\frac{k}{m}) \mu(\frac{k}{n}) \right) Y(mx). \end{aligned}$$

Proposition 2.18 The best approximation of a function $f(x) \in L^2[-\pi, \pi]$ by a superposition of the finite square waves

$$1, X(x), Y(x), X(2x), Y(2x), \cdots, X(Nx), \text{ and } Y(Nx) \tag{2.69}$$

is

$$f^{(N)}(x) = C_0 + \sum_{n=1}^{N} \left(C^{(N)}(n) X(nx) + D^{(N)}(n) Y(nx) \right), \tag{2.70}$$

where

$$\begin{aligned} C_0 &= \frac{1}{2\pi} \int_{-\pi}^{+\pi} f(x) dx, \\ C^{(N)}(n) &= \int_{-\pi}^{+\pi} f(x) h_n^{(N)}(x) dx, \\ D^{(N)}(n) &= \int_{-\pi}^{+\pi} f(x) g_n^{(N)}(x) dx. \end{aligned}$$

Proof. $f(x) \in L^2[-\pi, \pi]$ can be written as

$$f(x) = r_0 + \sum_{n=1}^{\infty} (r(n) E_n(x) + q(n) e_n(x)), \tag{2.71}$$

where

$$\begin{aligned} r_0 &= \tfrac{1}{2\pi} \textstyle\int_{-\pi}^{+\pi} f(x) dx, \\ r(n) &= \textstyle\int_{-\pi}^{+\pi} f(x) E_n(x) dx, \\ q(n) &= \textstyle\int_{-\pi}^{+\pi} f(x) e_n(x) dx. \end{aligned} \tag{2.72}$$

The best approximation to $f(x)$ out of the subspace spanned by

$$1, X(x), Y(x), X(2x), Y(2x), \cdots, X(Nx), \text{ and } Y(Nx)$$

is

$$
\begin{aligned}
f^{(N)}(x) &= r_0 + \sum_{n=1}^{N} \quad (r(n)E_n(x) + q(n)e_n(x)) \\
&= r_0 + \sum_{n=1}^{N} \quad [\, r(n)\frac{1}{\sqrt{\frac{\pi^3}{8}}}\frac{1}{\sqrt{\beta(n)}}\sum_{d|n}\mu(\frac{n}{d})A(\frac{n}{d})X(dx) \\
&\qquad +q(n)\frac{1}{\sqrt{\frac{\pi^3}{8}}}\frac{1}{\sqrt{\beta(n)}}\sum_{d|n}\mu(\frac{n}{d})B(\frac{n}{d})Y(dx)\,] \\
&= C_0 + \sum_{n=1}^{N} \quad \left(C^{(N)}(n)X(nx) + D^{(N)}(n)Y(nx)\right),
\end{aligned}
$$

where

$$
\begin{aligned}
& C^{(N)}(n) \\
= &\ \frac{1}{\sqrt{\frac{\pi^3}{8}}}\sum_{s=1}^{sn\leq N} \mu(s)A(s)r(sn)\frac{1}{\sqrt{\beta(sn)}} \\
= &\ \frac{1}{\sqrt{\frac{\pi^3}{8}}}\sum_{s=1}^{sn\leq N} \mu(s)A(s)\frac{1}{\sqrt{\beta(sn)}}\int_{-\pi}^{+\pi} f(x)E_{sn}(x)dx \\
= &\ \int_{-\pi}^{+\pi} f(x)\ \frac{1}{\sqrt{\frac{\pi^3}{8}}}\sum_{s=1}^{sn\leq N} \mu(s)A(s)\frac{1}{\sqrt{\beta(sn)}}E_{sn}(x)\ dx \\
= &\ \int_{-\pi}^{+\pi} f(x)\ \frac{1}{\frac{\pi^3}{8}}\sum_{s=1}^{sn\leq N} \mu(s)A(s)\frac{1}{\beta(sn)}\sum_{d|sn}\mu(\frac{sn}{d})A(\frac{sn}{d})X(dx)\ dx \\
= &\ \int_{-\pi}^{+\pi} f(x)\frac{1}{\frac{\pi^3}{8}}\sum_{m=1}^{N}(\sum_{m|k,n|k}^{1\leq k\leq N}\frac{1}{\beta(k)}A(\frac{k}{m})A(\frac{k}{n})\mu(\frac{k}{m})\mu(\frac{k}{n})\ )\ X(mx)\ dx \\
= &\ \int_{-\pi}^{+\pi} f(x)h_n^{(N)}(x)dx,
\end{aligned}
$$

and similarly

$$
D^{(N)}(n) = \int_{-\pi}^{+\pi} f(x)g_n^{(N)}(x)dx. \quad \square
$$

In fact, $\{h_n^{(N)}(x)\}_{n=1}^N$ are the biorthogonal functions of $\{X(nx)\}_{n=1}^N$ in the subspace spanned by $\{X(nx)\}_{n=1}^N$, and $\{g_n^{(N)}(x)\}_{n=1}^N$ are the biorthogonal functions of $\{Y(nx)\}_{n=1}^N$ in the subspace spanned by

$\{Y(nx)\}_{n=1}^{N}$.

Example. When $N = 3$, we have

$$\begin{aligned}
h_1^{(3)}(x) &= \frac{1}{\pi^3}(9X(x) + 3X(3x)),\\
h_2^{(3)}(x) &= \frac{1}{\pi^3}8X(2x),\\
h_3^{(3)}(x) &= \frac{1}{\pi^3}(3X(x) + 9X(3x)),\\
g_1^{(3)}(x) &= \frac{1}{\pi^3}(9Y(x) - 3Y(3x)),\\
g_2^{(3)}(x) &= \frac{1}{\pi^3}8Y(2x),\\
g_3^{(3)}(x) &= \frac{1}{\pi^3}(-3Y(x) + 9Y(3x)).
\end{aligned}$$

The best approximations of $\sin x$ and $\cos x$ are

$$\begin{aligned}
\sin^{(3)} x &= \frac{\pi}{\sqrt{\frac{\pi^3}{8}}}\sum_{n=1}^{3}\frac{1}{\sqrt{\beta(n)}}\mu(n)B(n)e_n(x)\\
&= \frac{1}{\pi^2}(9Y(x) - 3Y(3x)),\\
\cos^{(3)} x &= \frac{\pi}{\sqrt{\frac{\pi^3}{8}}}\sum_{n=1}^{3}\frac{1}{\sqrt{\beta(n)}}\mu(n)A(n)E_n(x)\\
&= \frac{1}{\pi^2}(9X(x) + 3X(3x)).
\end{aligned}$$

Proposition 2.19 Let N_1 and N_2 be two positive integers and $f(x) \in L^2[-\pi, \pi]$. Then $f(x)$'s best approximation by a linear combination of 1, $\{X(dx)\}_{d|N_1}$ and $\{X(dx)\}_{d|N_2}$, is

$$f^{[N_1,N_2]}(x) = C_0 + \sum_{d|N_1} C^{[N_1]}(d)X(dx) + \sum_{d|N_2} D^{[N_2]}(d)Y(dx). \tag{2.73}$$

The coefficients in (2.73) are determined by

$$\begin{aligned}
C_0 &= \frac{1}{2\pi}\int_{-\pi}^{+\pi} f(x)dx,\\
C^{[N_1]}(d) &= \int_{-\pi}^{+\pi} f(x)h_d^{[N_1]}(x)dx,\\
D^{[N_2]}(d) &= \int_{-\pi}^{+\pi} f(x)g_d^{[N_2]}(x)dx,
\end{aligned}$$

where

$$h_d^{[N_1]}(x) = \frac{1}{\frac{\pi^3}{8}} \sum_{m|N_1} \left(\sum_{m|k,d|k}^{k|N_1} \frac{1}{\beta(k)} A(\frac{k}{m}) A(\frac{k}{d}) \mu(\frac{k}{m}) \mu(\frac{k}{d}) \right) X(mx),$$

$$g_d^{[N_2]}(x) = \frac{1}{\frac{\pi^3}{8}} \sum_{m|N_2} \left(\sum_{m|k,d|k}^{k|N_2} \frac{1}{\beta(k)} B(\frac{k}{m}) B(\frac{k}{d}) \mu(\frac{k}{m}) \mu(\frac{k}{d}) \right) Y(mx).$$

Proof. It is plain that $f(x)$'s best approximation in this case is

$$f^{[N_1,N_2]}(x) = r_0 + \sum_{d|N_1} r(d)E_d(x) + \sum_{d|N_2} q(d)e_d(x). \tag{2.74}$$

The rest of the proof is similar to that of proposition 2.18. □

Example. $N_1 = N_2 = 6$.

The best approximations of $\sin x, \cos x$ by a superposition of the square waves 1, $Y(x), Y(2x), Y(3x), Y(6x)$ and $X(x), X(2x), X(3x)$, $X(6x)$ are

$$\begin{aligned} \sin^{[6,6]} x &= \frac{\pi}{\sqrt{\frac{\pi^3}{8}}} \sum_{d|6} \frac{1}{\sqrt{\beta(d)}} \mu(d)B(d)e_d(x) \\ &= \frac{1}{\pi^2}(9Y(x) - 3Y(3x)), \\ \cos^{[6,6]} x &= \frac{\pi}{\sqrt{\frac{\pi^3}{8}}} \sum_{d|6} \frac{1}{\sqrt{\beta(d)}} \mu(d)A(d)E_d(x) \\ &= \frac{1}{\pi^2}(9X(x) + 3X(3x)). \end{aligned}$$

Remark. In applications we have to express a signal as a superposition of finite square waves rather than infinite square waves, since the number of square waves is finite after all, and therefore the best approximation is an important practical problem.

2.4 An Example of Applications

The idea that a signal can be expressed as a series of square waves with different frequencies will bring us many completely new designs in electronics, because once a signal is associated with a square wave series, one can record, restore and transmit the square wave coefficients of the signal instead of the signal itself, and if necessary one can recover the signal from its square wave coefficients. A series of techniques based on sine-cosine functions, such as filtering and multiplexing, can be translated into those based on square waves. The technical aspect of common waveforms will be discussed in details in Chapter 5. Here we only give a small example: Fourier analysis based on square waves,

Fourier analysis of an analog signal is to deal with integrations of products of the signal and sine-cosine functions. The main difficulty of designing Fourier analyzer is due to accurate multipliers for two continuous functions. With the square wave expansions (2.18) and (2.19) of sine and cosine functions, simple multipliers can do this, because

$$\begin{aligned} a(n) &= \int_{-\pi}^{+\pi} f(x)\cos nx dx \\ &= \sum_{m=1}^{\infty} \mu(m)A(m)\int_{-\pi}^{+\pi} f(x)X(mnx)dx, \end{aligned} \qquad (2.75)$$

and

$$\begin{aligned} b(n) &= \int_{-\pi}^{+\pi} f(x)\sin nx dx \\ &= \sum_{m=1}^{\infty} \mu(m)B(m)\int_{-\pi}^{+\pi} f(x)Y(mnx)dx. \end{aligned} \qquad (2.76)$$

The schematic diagram of Fourier analyzer based on (2.75) and (2.76) is shown in Figure 2.6. Obviously multipliers for a signal and a sine-cosine functions are replaced by multipliers for a signal and a square wave. In practice, ∞ in (2.75) and (2.76) should be considered as a sufficiently large number.

By the way, AFT (arithmetic Fourier analysis) presented by Tufts and Sadasive has a relation to Möbius function as well [6].

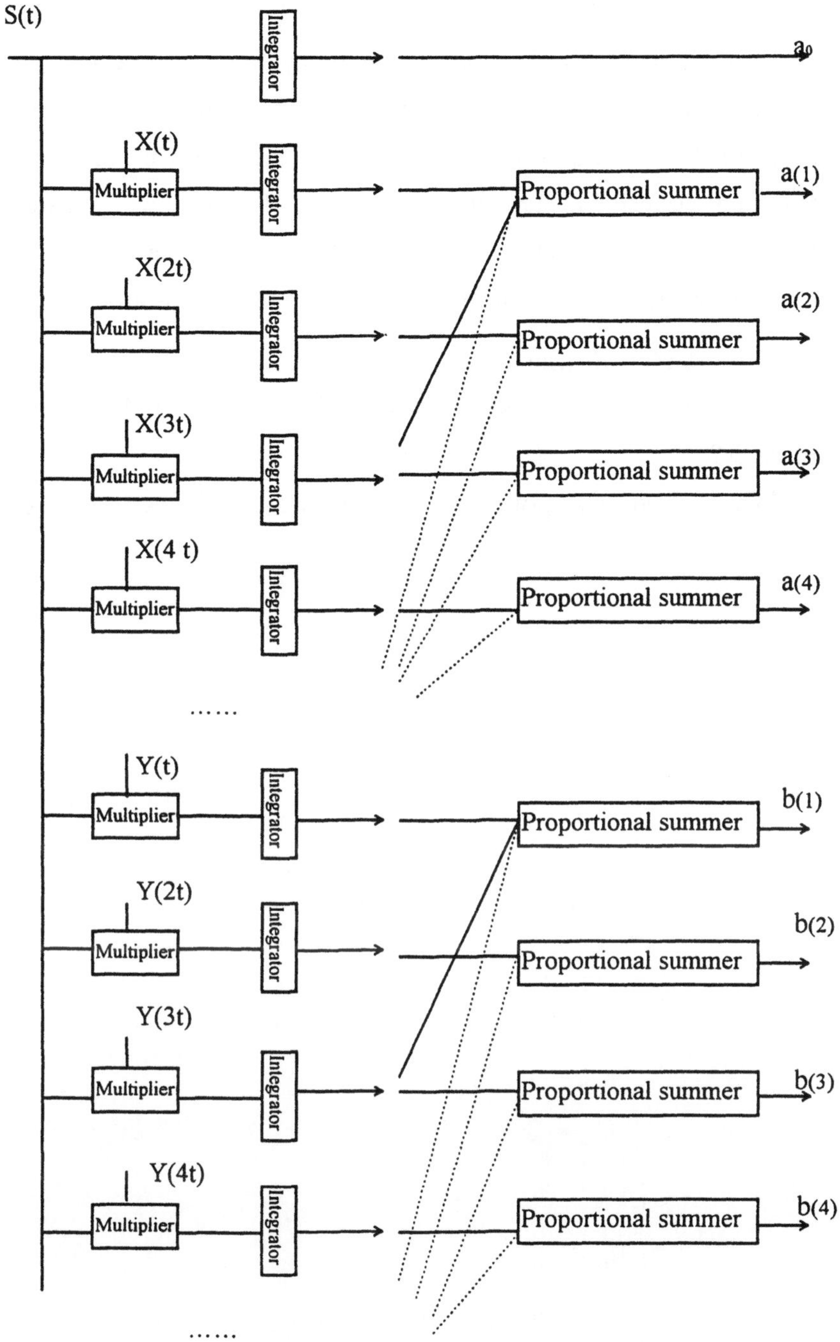

Figure 2.6: Schematic diagram of Fourier analysis based on square waves.

SUMMARY

From the practical viewpoint, one can say that a signal can be considered as a superposition of square waves with different frequencies and a signal can be decomposed into a square wave series by means of the biorthogonal functions. The square wave coefficients and Fourier coefficients are related via Dirichlet multiplication. These results form a theoretical foundation for the technique of square wave analysis.

Unsolved Problems:

1. Is the square wave system a basis of $L^2[-\pi,\pi]$? Is it an unconditional basis?

2. What about the pointwise convergence of the series (2.18) and (2.19)?

Chapter 3

TRIANGULAR WAVE ANALYSIS AND TRAPEZOIDAL WAVE ANALYSIS

Besides square wave, triangular wave and trapezoidal wave are common waveforms in modern electronics as well. This chapter is devoted to triangular wave analysis and trapezoidal wave analysis. Main questions include:

1. How to express a signal as a triangular wave series or a trapezoidal wave series?

2. What are the relations of triangular wave series, trapezoidal wave series, and Fourier series?

3. If a signal is known only at some discrete points, how to represent the signal approximately as a summation of finite triangular waves or trapezoidal waves?

In order to answer the above questions we shall consider the following items:

1. WASCMFC functions and practical examples;
2. WASCMFC function basis, biorthogonal basis and orthonormal basis;
3. Basis and coordinate transforms;
4. Discrete triangular wave transform and trapezoidal wave transform.

3.1 WASCMFC Functions and Practical Examples

We introduce the concept of WASCMFC functions firstly, and then show that triangular wave and trapezoidal wave are practical examples. In fact all our discussion in this chapter are based on this concept.

Definition. If a real-valued arithmetical function $B(n)$ is absolutely summable, i.e., $\sum_{n=1}^{\infty} \mid B(n) \mid < \infty$, and completely multiplicative, then the function $Y(x)$ defined on $(-\infty, +\infty)$ by

$$Y(x) = \sum_{n=1}^{\infty} B(n)\sin(nx) \tag{3.1}$$

is called an odd function **W**ith **A**bsolutely **S**ummable and **C**ompletely **M**ultiplicative **F**ourier **C**oefficients, or an odd WASCMFC function.

Of course $Y(x)$ is a real-valued continuous function with period 2π.

We write Y for the set of all odd WASCMFC functions, that is,

$$Y = \{\ Y(x) = \sum_{n=1}^{\infty} B(n)\sin(nx) : \quad B(n) \text{ is absolutely summable and completely multiplicative } \}.$$

Similarly, we write X for the set of all even WASCMFC functions, that is,

$$X = \{\ X(x) = \sum_{n=1}^{\infty} A(n)\cos(nx) : \quad A(n) \text{ is absolutely summable and completely multiplicative } \}.$$

The following four common electrical waveforms are practical examples of WASCMFC functions. See Figure 3.1.

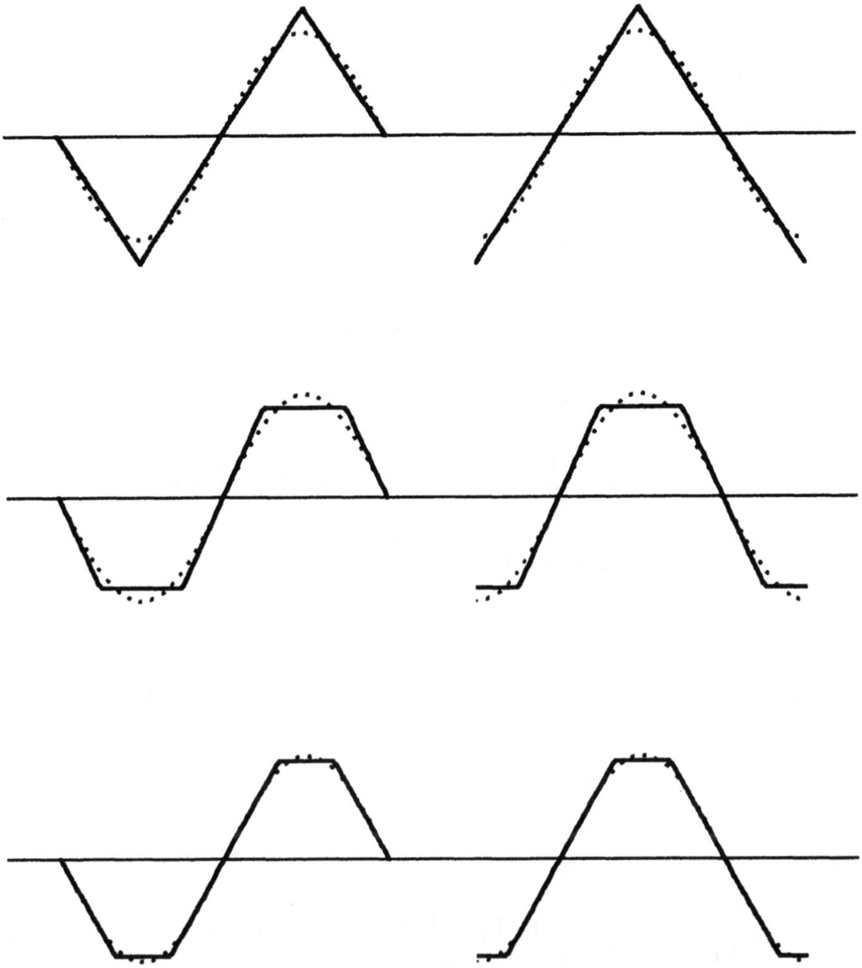

Figure 3.1. The triangular waves, the I-trapezoidal waves and the II-trapezoidal waves.

$$\begin{pmatrix} Y_{\mathrm{tri}}(x) & X_{\mathrm{tri}}(x) \\ Y_{\mathrm{Itra}}(x) & X_{\mathrm{Itra}}(x) \\ Y_{\mathrm{IItra}}(x) & X_{\mathrm{IItra}}(x) \end{pmatrix}$$

1. $\sin x$ and $\cos x$

By the definitions, the functions $\sin x$ and $\cos x$ are odd and even WASCMFC functions trivially, but we shall pay attention to other waveforms.

2. The triangular waves $Y_{\text{tri}}(x)$ and $X_{\text{tri}}(x)$.

They are periodic functions with period 2π defined by

$$Y_{\text{tri}}(x) = \begin{cases} \frac{\pi}{4}x & -\frac{\pi}{2} \le x < \frac{\pi}{2}, \\ \frac{\pi^2 - \pi x}{4} & \frac{\pi}{2} \le x < \frac{3\pi}{2}, \end{cases} \tag{3.2}$$

$$X_{\text{tri}}(x) = Y_{\text{tri}}(x + \pi/2). \tag{3.3}$$

Their Fourier series are [3]

$$\begin{aligned} Y_{\text{tri}}(x) &= \sin x - \frac{1}{3^2}\sin(3x) + \frac{1}{5^2}\sin(5x) - \frac{1}{7^2}\sin(7x) + \cdots \\ &= \sum_{n=1}^{\infty} B_{\text{tri}}(n)\sin(nx), \end{aligned}$$

$$\begin{aligned} X_{\text{tri}}(x) &= \cos x + \frac{1}{3^2}\cos(3x) + \frac{1}{5^2}\cos(5x) + \frac{1}{7^2}\cos(7x) + \cdots \\ &= \sum_{n=1}^{\infty} A_{\text{tri}}(n)\cos(nx), \end{aligned}$$

where

$$B_{\text{tri}}(n) = \begin{cases} (-1)^{(n-1)/2}\frac{1}{n^2} & n = 1, 3, 5, \cdots, \\ 0 & n = 2, 4, 6, \cdots, \end{cases} \tag{3.4}$$

$$A_{\text{tri}}(n) = \begin{cases} \frac{1}{n^2} & n = 1, 3, 5, \cdots, \\ 0 & n = 2, 4, 6, \cdots, \end{cases} \tag{3.5}$$

are absolutely summable and completely multiplicative.

3. The I-trapezoidal waves $Y_{\text{Itra}}(x)$ and $X_{\text{Itra}}(x)$.
They are periodic functions with period 2π defined by

$$Y_{\text{Itra}}(x) = \begin{cases} -\frac{\sqrt{2}\pi^2}{16} & x \in [-\frac{3\pi}{4}, -\frac{\pi}{4}], \\ \frac{\sqrt{2}\pi}{4}x & x \in [-\frac{\pi}{4}, \frac{\pi}{4}], \\ \frac{\sqrt{2}\pi^2}{16} & x \in [\frac{\pi}{4}, \frac{3\pi}{4}], \\ \frac{\sqrt{2}\pi}{4}(\pi - x) & x \in [\frac{3\pi}{4}, \frac{5\pi}{4}], \end{cases} \tag{3.6}$$

$$X_{\text{Itra}}(x) = Y_{\text{Itra}}(x + \pi/2). \tag{3.7}$$

Their Fourier series are [3]

$$\begin{aligned} Y_{\text{Itra}}(x) &= \sin x + \frac{1}{3^2}\sin(3x) - \frac{1}{5^2}\sin(5x) - \frac{1}{7^2}\sin(7x) + \cdots \\ &= \sum_{n=1}^{\infty} B_{\text{Itra}}(n)\sin(nx), \end{aligned}$$

$$\begin{aligned} X_{\text{Itra}}(x) &= \cos x - \frac{1}{3^2}\cos(3x) - \frac{1}{5^2}\cos(5x) + \frac{1}{7^2}\cos(7x) + \cdots \\ &= \sum_{n=1}^{\infty} A_{\text{Itra}}(n)\cos(nx), \end{aligned}$$

where

$$B_{\text{Itra}}(n) = \begin{cases} \frac{1}{n^2} & n = 8l+1, \\ 0 & n = 8l+2, \\ \frac{1}{n^2} & n = 8l+3, \\ 0 & n = 8l+4, \\ -\frac{1}{n^2} & n = 8l+5, \\ 0 & n = 8l+6, \\ -\frac{1}{n^2} & n = 8l+7, \\ 0 & n = 8l+8, \end{cases} \qquad (l = 0, 1, 2, 3, \cdots) \tag{3.8}$$

$$A_{\mathrm{Itra}}(n) = \begin{cases} \frac{1}{n^2} & n = 8l+1, \\ 0 & n = 8l+2, \\ -\frac{1}{n^2} & n = 8l+3, \\ 0 & n = 8l+4, \\ -\frac{1}{n^2} & n = 8l+5, \\ 0 & n = 8l+6, \\ \frac{1}{n^2} & n = 8l+7, \\ 0 & n = 8l+8, \end{cases} \qquad (l = 0,1,2,3,\cdots) \tag{3.9}$$

are absolutely summable and completely multiplicative.

4. The II-trapezoidal waves $Y_{\mathrm{IItra}}(x)$ and $X_{\mathrm{IItra}}(x)$.

They are periodic functions with period 2π defined by

$$Y_{\mathrm{IItra}}(x) = \begin{cases} -\frac{\pi^2}{6\sqrt{3}} & x \in [-\frac{2\pi}{3}, -\frac{\pi}{3}], \\ \frac{\pi}{2\sqrt{3}}x & x \in [-\frac{\pi}{3}, \frac{\pi}{3}], \\ \frac{\pi^2}{6\sqrt{3}} & x \in [\frac{\pi}{3}, \frac{2\pi}{3}], \\ \frac{\pi}{2\sqrt{3}}(\pi - x) & x \in [\frac{2\pi}{3}, \frac{4\pi}{3}], \end{cases} \tag{3.10}$$

$$X_{\mathrm{IItra}}(x) = Y_{\mathrm{IItra}}(x + \pi/2). \tag{3.11}$$

Their Fourier series are [3]

$$\begin{aligned} Y_{\mathrm{IItra}}(x) &= \sin x - \frac{1}{5^2}\sin(5x) + \frac{1}{7^2}\sin(7x) - \frac{1}{11^2}\sin(11x) + \cdots \\ &= \sum_{n=1}^{\infty} B_{\mathrm{IItra}}(n)\sin(nx), \end{aligned}$$

$$\begin{aligned} X_{\mathrm{IItra}}(x) &= \cos x - \frac{1}{5^2}\cos(5x) - \frac{1}{7^2}\cos(7x) + \frac{1}{11^2}\cos(11x) + \cdots \\ &= \sum_{n=1}^{\infty} A_{\mathrm{IItra}}(n)\cos(nx), \end{aligned}$$

where

$$B_{\text{IItra}}(n) = \begin{cases} \frac{1}{n^2} & n = 6l+1, \\ 0 & n = 6l+2, \\ 0 & n = 6l+3, \\ 0 & n = 6l+4, \\ -\frac{1}{n^2} & n = 6l+5, \\ 0 & n = 6l+6, \end{cases} \qquad (l = 0, 1, 2, 3, \cdots) \tag{3.12}$$

$$A_{\text{IItra}}(n) = \begin{cases} \frac{1}{n^2} & n = 12l+1, \\ 0 & n = 12l+2, \\ 0 & n = 12l+3, \\ 0 & n = 12l+4, \\ -\frac{1}{n^2} & n = 12l+5, \\ 0 & n = 12l+6, \\ -\frac{1}{n^2} & n = 12l+7, \\ 0 & n = 12l+8, \\ 0 & n = 12l+9, \\ 0 & n = 12l+10, \\ +\frac{1}{n^2} & n = 12l+11, \\ 0 & n = 12l+12, \end{cases} \qquad (l = 0, 1, 2, 3, \cdots) \tag{3.13}$$

are absolutely summable and completely multiplicative.

Proposition 3.1 Let $A(n)$ and $B(n)$ be the n-th Fourier coefficients of $X(x) \in X$ and $Y(x) \in Y$, respectively. The series

$$\cos(x) = \sum_{n=1}^{\infty} \mu(n) A(n) X(nx), \tag{3.14}$$

and

$$\sin(x) = \sum_{n=1}^{\infty} \mu(n) B(n) Y(nx). \tag{3.15}$$

converge absolutely and uniformly in $(-\infty, +\infty)$.

Proof. Let's consider the series (3.14) first. We have

$$\sum_{n=1}^{\infty} \mu(n) A(n) X(nx)$$

$$= \sum_{n=1}^{\infty} \mu(n)A(n) \sum_{m=1}^{\infty} A(m)\cos(mnx) \quad (3.16)$$

$$= \sum_{k=1}^{\infty} \sum_{d|k} \mu(d)A(k)\cos kx \quad (3.17)$$

$$= \cos x,$$

where $d \mid k$ means that d is a factor of k.

From (3.16) to (3.17) one changes the summation order. The reason for this is that the series (3.16) converges absolutely. In fact we have

$$\sum_{n=1}^{\infty} | \mu(n)A(n) | \sum_{m=1}^{\infty} | A(m)\cos(mnx) |$$
$$\leq \sum_{n=1}^{\infty} | A(n) | \sum_{m=1}^{\infty} | A(m) |$$
$$= \left(\sum_{n=1}^{\infty} | A(n) | \right)^2$$
$$< +\infty.$$

By Weierstrass M-test the series (3.14) converges absolutely and uniformly in $(-\infty, +\infty)$.

The case of the series (3.15) is similar. □

The Weierstrass approximation theorem asserts that any real-valued continuous function defined on $(-\infty, +\infty)$ with period 2π may be approximated uniformly by trigonometric polynomials. By Proposition 3.1, we obtain a generalization of this theorem easily.

Proposition 3.2 Suppose $X(x) \in X$ and $Y(x) \in Y$ are given functions. Then any real-valued continuous function defined on $(-\infty, +\infty)$ with period 2π may be approximated uniformly by linear combinations of the form

$$C_0 + \sum_{n=1}^{N} (C(n)X(nx) + D(n)Y(nx)), \quad (3.18)$$

where $C_0, C(n)$, and $D(n)$ are real numbers and N is a positive integer.

In case $X(x) = \cos x$ and $Y(x) = \sin x$, Proposition 3.2 becomes the Weierstrass approximation theorem.

3.2 WASCMFC Function Basis, Biorthogonal Basis and Orthonormalized Basis

In this section, we suppose that $X(x) \in X$ and $Y(x) \in Y$ are two given WASCMFC functions, and that $A(n)$ and $B(n)$ are their Fourier coefficients, respectively.

We shall show that the function system

$$1, X(x), Y(x), X(2x), Y(2x), \cdots, X(nx), Y(nx), \cdots \tag{3.19}$$

is an unconditional basis of $L^2[-\pi, \pi]$, and present its biorthogonal basis and orthonormalized basis explicitly.

Definition. For $n = 1, 2, 3, \cdots$, functions $h_n(x)$ and $g_n(x)$ are defined by

$$h_n(x) = \frac{1}{\pi} \sum_{d|n} \mu(\frac{n}{d}) A(\frac{n}{d}) \cos dx, \tag{3.20}$$

and

$$g_n(x) = \frac{1}{\pi} \sum_{d|n} \mu(\frac{n}{d}) B(\frac{n}{d}) \sin dx, \tag{3.21}$$

respectively.

Proposition 3.3 Any function $f(x) \in L^2[-\pi, \pi]$ has an unconditionally convergent series

$$f(x) = C_0 + \sum_{n=1}^{\infty} [C(n)X(nx) + D(n)Y(nx)], \tag{3.22}$$

where the coefficients can be determined by

$$\begin{aligned} C_0 &= \tfrac{1}{2\pi} \textstyle\int_{-\pi}^{+\pi} f(x)dx, \\ C(n) &= \textstyle\int_{-\pi}^{+\pi} f(x)h_n(x)dx, \\ D(n) &= \textstyle\int_{-\pi}^{+\pi} f(x)g_n(x)dx. \end{aligned} \tag{3.23}$$

Proof. It is well-known that any function $f(x) \in L^2[-\pi, \pi]$ has a Fourier series:

$$f(x) = a_0 + \sum_{n=1}^{\infty} [a(n) \cos nx + b(n) \sin nx], \tag{3.24}$$

where the Fourier coefficients are determined by

$$
\begin{aligned}
a_0 &= \frac{1}{2\pi}\int_{-\pi}^{+\pi} f(x)dx,\\
a(n) &= \frac{1}{\pi}\int_{-\pi}^{+\pi} f(x)\cos nx dx,\\
b(n) &= \frac{1}{\pi}\int_{-\pi}^{+\pi} f(x)\sin nx dx.
\end{aligned}
$$

Obviously, the series (3.14) and (3.15) hold in the sense of the norm of $L^2[-\pi,\pi]$. Putting them into (3.24), one obtains that

$$
\begin{aligned}
f(x) &= a_0 + \sum_{n=1}^{\infty} a(n)\sum_{l=1}^{\infty} A(l)\mu(l)X(nlx)\\
&\quad + \sum_{n=1}^{\infty} b(n)\sum_{l=1}^{\infty} B(l)\mu(l)Y(nlx) \qquad (3.25)\\
&= a_0 + \sum_{n=1}^{\infty}\Big(\sum_{d|n} a(d)\mu(\frac{n}{d})A(\frac{n}{d})\Big)X(nx)\\
&\quad + \sum_{n=1}^{\infty}\Big(\sum_{d|n} b(d)\mu(\frac{n}{d})B(\frac{n}{d})\Big)Y(nx) \qquad (3.26)\\
&= C_0 + \sum_{n=1}^{\infty}[C(n)X(nx) + D(n)Y(nx)],
\end{aligned}
$$

where

$$
\begin{aligned}
C_0 &= a_0,\\
C(n) &= \sum_{d|n}\mu(\frac{n}{d})A(\frac{n}{d})a(d),\\
D(n) &= \sum_{d|n}\mu(\frac{n}{d})B(\frac{n}{d})b(d),
\end{aligned}
$$

or,

$$
\begin{aligned}
C_0 &= \frac{1}{2\pi}\int_{-\pi}^{+\pi} f(x)dx,\\
C(n) &= \int_{-\pi}^{+\pi} f(x)h_n(x)dx,\\
D(n) &= \int_{-\pi}^{+\pi} f(x)g_n(x)dx.
\end{aligned}
$$

It should be pointed out that from (3.25) to (3.26) the summation order is changed. This does not matter, because the series (3.25) converges unconditionally in $L^2[-\pi, \pi]$. The reason for this is shown below.

There are two series in (3.25), and let us consider one of the two, for the cases are the same. The following series

$$\begin{aligned}
&\sum_{n=1}^{\infty} a(n) \sum_{l=1}^{\infty} A(l)\mu(l)X(nlx) \\
= &\sum_{n=1}^{\infty} a(n) \sum_{l=1}^{\infty} A(l)\mu(l) \sum_{m=1}^{\infty} A(m)\cos(nlmx) \\
= &\sum_{k=1}^{\infty}\Big(\sum_{nlm=k} a(n)A(l)\mu(l)A(m) \Big) \cos(kx)
\end{aligned}$$

are unconditionally convergent in $L^2[-\pi, \pi]$, because the series

$$\sum_{k=1}^{\infty}\Big(\sum_{nlm=k} | a(n)A(l)\mu(l)A(m) |\Big) \cos(kx) \tag{3.27}$$

is convergent, the norm of which is

$$\begin{aligned}
&\Big\| \sum_{k=1}^{\infty}\Big(\sum_{nlm=k} | a(n)A(l)\mu(l)A(m) |\Big) \cos(kx) \Big\| \\
= &\Big\| \sum_{m=1}^{\infty} | A(m) | \sum_{l=1}^{\infty} | A(l)\mu(l) | \sum_{n=1}^{\infty} | a(n) | \cos(nlmx) \Big\| \\
\leq &\sum_{m=1}^{\infty} | A(m) | \sum_{l=1}^{\infty} | A(l)\mu(l) | \Big\| \sum_{n=1}^{\infty} | a(n) | \cos(nlmx) \Big\| \\
= &\sum_{m=1}^{\infty} | A(m) | \sum_{l=1}^{\infty} | A(l)\mu(l) | \sqrt{\pi \sum_{n=1}^{\infty} a^2(n)} \\
\leq &\sqrt{\pi \sum_{n=1}^{\infty} a^2(n)} \left(\sum_{m=1}^{\infty} | A(m) | \right)^2 \\
< &\infty. \quad \Box
\end{aligned}$$

Proposition 3.4 If $f(x) \in L^2[-\pi, \pi]$ has a WASCMFC function series:

$$f(x) = C_0 + \sum_{n=1}^{\infty}(C(n)X(nx) + D(n)Y(nx)), \tag{3.28}$$

then all the coefficients $C_0, C(n)$, and $D(n)$ are unique.

Proof. Suppose there exist another group of coefficients such that

$$f(x) = C_0' + \sum_{n=1}^{\infty}(C'(n)X(nx) + D'(n)Y(nx)). \tag{3.29}$$

Subtracting (3.29) from (3.28), one obtains

$$(C_0 - C_0') + \sum_{n=1}^{\infty}[(C(n) - C'(n))X(nx) + (D(n) - D'(n))Y(nx)] = 0.$$

Substituting the Fourier series of $X(nx)$ and $Y(nx)$ for $X(nx)$ and $Y(nx)$, one obtains that

$$a_0 + \sum_{n=1}^{\infty}(a(n)\cos nx + b(n)\sin nx) = 0,$$

where

$$\begin{aligned} a_0 &= C_0 - C_0' = 0, \\ a(n) &= \sum_{d|n} A(\frac{n}{d})(C(d) - C'(d)) = 0, \\ b(n) &= \sum_{d|n} B(\frac{n}{d})(D(d) - D'(d)) = 0. \end{aligned}$$

Therefore we have

$$\begin{aligned} C_0 - C_0' &= 0, \\ C(n) - C'(n) &= \sum_{d|n} 0 \bullet \mu(\frac{n}{d})A(\frac{n}{d}) = 0, \\ D(n) - D'(n) &= \sum_{d|n} 0 \bullet \mu(\frac{n}{d})B(\frac{n}{d}) = 0. \end{aligned}$$

In other words, $C_0, C(n)$, and $D(n)$ are unique. □

By Proposition 3.3 and proposition 3.4, we have the following proposition.

Proposition 3.5 The function system (3.19) is an unconditional basis of $L^2[-\pi,\pi]$. (We call it a WASCMFC function basis.)

Consequence: The function system

$$\frac{1}{2\pi}, h_1(x), g_1(x), h_2(x), g_2(x), \cdots, h_n(x), g_n(x), \cdots, \tag{3.30}$$

is the biorthogonal basis of the WASCMFC function basis (3.19).

Generally, the WASCMFC function basis (3.19) is nonorthogonal unless $X(x) = \cos x$ and $Y(x) = \sin x$. It can be orthonormalized by means of the Gram-Schmidt orthogonalization process.

Definition. For $n = 1, 2, 3, \cdots$, the functions $E_n(x)$ and $e_n(x)$ are defined by

$$E_n(x) = \frac{1}{\| X(x) \|} \frac{1}{\sqrt{\alpha(n)}} \sum_{d|n} \mu(\frac{n}{d}) A(\frac{n}{d}) X(dx), \tag{3.31}$$

and

$$e_n(x) = \frac{1}{\| Y(x) \|} \frac{1}{\sqrt{\beta(n)}} \sum_{d|n} \mu(\frac{n}{d}) B(\frac{n}{d}) Y(dx), \tag{3.32}$$

where $\alpha(n)$ and $\beta(n)$ are defined by

$$\alpha(n) = \sum_{d|n} \mu(d) A^2(d) \tag{3.33}$$

and

$$\beta(n) = \sum_{d|n} \mu(d) B^2(d). \tag{3.34}$$

Proposition 3.6 The function system

$$\frac{1}{\sqrt{2\pi}}, E_1(x), e_1(x), E_2(x), e_2(x), \cdots, E_n(x), e_n(x), \cdots, \tag{3.35}$$

is the orthonormalized basis of the WASCMFC function basis (3.19).

Proof. This proof is similar to that of Proposition 2.16.

Thus for the given WASCMFC functions $X(x) \in X$ and $Y(x) \in Y$, we obtain three related bases: the WASCMFC function basis (3.19), the biorthogonal basis (3.30), and the orthonormal basis (3.35). Next we shall consider their relations.

3.3 Basis and Coordinate Transforms

For convenience, we shall consider basis transforms and coordinate transforms in matrix notation only in the odd function subspace of $L^2[-\pi, +\pi]$, i.e. $\overline{\text{span}\{\sin nx\}_{n=1}^{\infty}}$. The conclusion in the even function subspace is the same.

This section is divided into two parts.

1. Suppose that $Y(x)$ is a given odd WASCMFC function, i.e., $Y(x) \in Y$, and that its nth Fourier coefficient is $B(n)$.

There are four related bases:

1) the WASCMFC function basis

$$Y(x), Y(2x), Y(3x), \cdots, Y(nx), \cdots, \tag{3.36}$$

2) the biorthogonal basis

$$g_1(x), g_2(x), g_3(x), \cdots, g_n(x), \cdots, \tag{3.37}$$

3) the orthonormal basis

$$e_1(x), e_2(x), e_3(x), \cdots, e_n(x), \cdots, \tag{3.38}$$

4) the sine function basis

$$\sin(x), \sin(2x), \sin(3x), \cdots, \sin(nx), \cdots. \tag{3.39}$$

It is well-known that every two bases are related by two mutually inverse matrices, as shown in Figure 3.2. Now let us determine the elements of these matrices.

Formula 3.1

$$(Y(kx))_{\infty}^{k=1} \underset{W^{-1}}{\overset{W}{\rightleftarrows}} (\sin kx)_{\infty}^{k=1}. \tag{3.40}$$

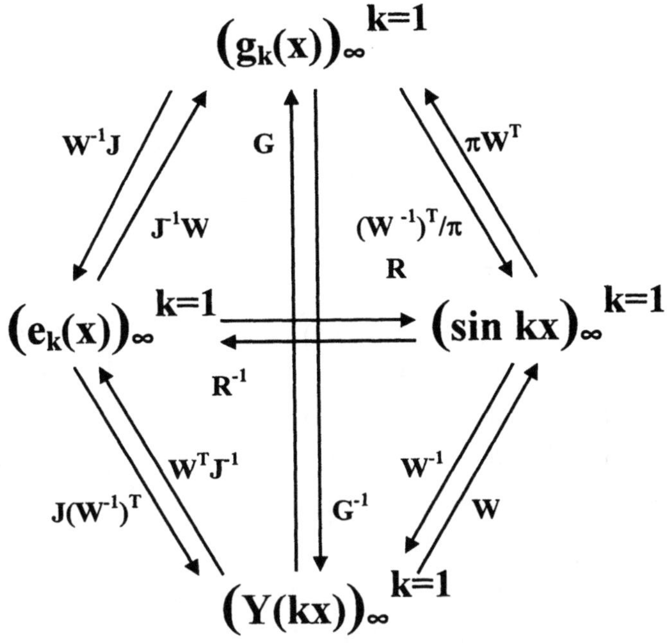

Figure 3.2: The relations of the four bases.

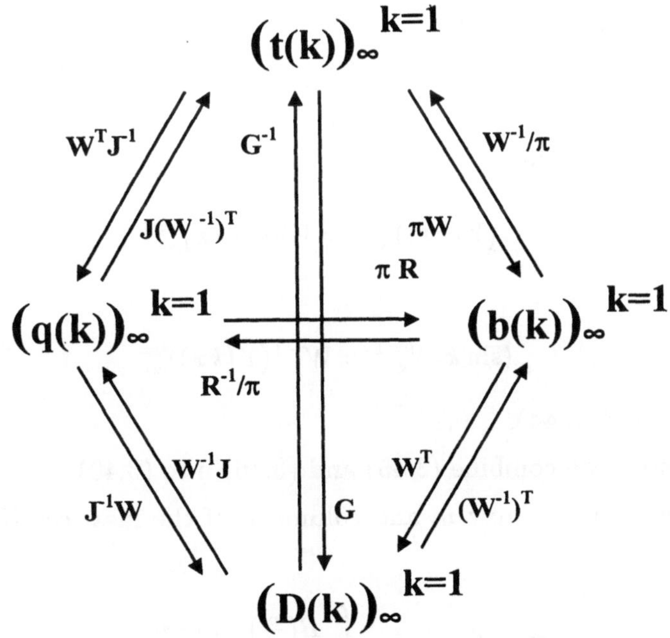

Figure 3.3: The relations of the four coordinates.

From (3.1) and (3.15), it follows that

$$Y(mx) = \sum_{n=1}^{\infty} B(n) \sin mnx, \tag{3.41}$$

$$\sin mx = \sum_{n=1}^{\infty} \mu(n) B(n) Y(mnx), \tag{3.42}$$

for $m = 1, 2, 3, \cdots$.
In matrix notation, we have

$$\begin{pmatrix} Y(x) \\ Y(2x) \\ \vdots \\ Y(kx) \\ \vdots \end{pmatrix} = \begin{pmatrix} B(1) & B(2) & B(3) & B(4) & \cdots \\ 0 & B(1) & 0 & B(2) & \cdots \\ 0 & 0 & B(1) & 0 & \cdots \\ 0 & 0 & 0 & B(1) & \cdots \\ \vdots & \vdots & \vdots & \vdots & \vdots \end{pmatrix} \begin{pmatrix} \sin(x) \\ \sin(2x) \\ \vdots \\ \sin(kx) \\ \vdots \end{pmatrix} \tag{3.43}$$

$$\begin{pmatrix} \sin(x) \\ \sin(2x) \\ \vdots \\ \sin(kx) \\ \vdots \end{pmatrix} = \begin{pmatrix} B(1)\mu(1) & B(2)\mu(2) & B(3)\mu(3) & \cdots \\ 0 & B(1)\mu(1) & 0 & \cdots \\ 0 & 0 & B(1)\mu(1) & \cdots \\ 0 & 0 & 0 & \cdots \\ \vdots & \vdots & \vdots & \vdots \end{pmatrix} \begin{pmatrix} Y(x) \\ Y(2x) \\ \vdots \\ Y(kx) \\ \vdots \end{pmatrix}. \tag{3.44}$$

For short, we write

$$(Y(kx))_{\infty}^{k=1} = W(\sin kx)_{\infty}^{k=1} \tag{3.45}$$

and

$$(\sin kx)_{\infty}^{k=1} = W^{-1}(Y(kx))_{\infty}^{k=1} \tag{3.46}$$

for (3.43) and (3.44).
Furthermore, we combine (3.45) and (3.46) into (3.40).

The elements in row m and column n of the matrices W and W^{-1} are

$$W(m,n) = \begin{cases} B(\frac{n}{m}) & m \mid n, \\ 0 & m \nmid n, \end{cases} \tag{3.47}$$

$$W^{-1}(m,n) = \begin{cases} \mu(\frac{n}{m})B(\frac{n}{m}) & m \mid n, \\ 0 & m \nmid n. \end{cases} \tag{3.48}$$

Similar notations will be used in the following formulas without explanations.

Formula 3.2

$$(g_k(x))_{\infty}^{k=1} \underset{\pi W^T}{\overset{(W^{-1})^T/\pi}{\rightleftarrows}} (\sin kx)_{\infty}^{k=1}. \tag{3.49}$$

From the definition of $g_n(x)$ (3.21) it follows easily that

$$(g_k(x))_{\infty}^{k=1} = \frac{1}{\pi}(W^{-1})^T(\sin kx)_{\infty}^{k=1}, \tag{3.50}$$

where T denotes transpose. Accordingly we have

$$(\sin kx)_{\infty}^{k=1} = \pi W^T(g_k(x))_{\infty}^{k=1}. \tag{3.51}$$

The elements of the matrices $(W^{-1})^T$ and W^T are

$$(W^{-1})^T(m,n) = W^{-1}(n,m) = \begin{cases} \mu(\frac{m}{n})B(\frac{m}{n}) & n \mid m, \\ 0 & n \nmid m, \end{cases} \tag{3.52}$$

$$W^T(m,n) = W(n,m) = \begin{cases} B(\frac{m}{n}) & n \mid m, \\ 0 & n \nmid m. \end{cases} \tag{3.53}$$

Formula 3.3

$$(e_k(x))_{\infty}^{k=1} \underset{W^T J^{-1}}{\overset{J(W^{-1})^T}{\rightleftarrows}} (Y(kx))_{\infty}^{k=1}. \tag{3.54}$$

Here we introduce two mutually inverse diagonal matrices J and J^{-1}, whose elements are

$$J(m,n) = \begin{cases} \frac{1}{\|Y(x)\|}\frac{1}{\sqrt{\beta(n)}} & m = n, \\ 0 & m \neq n, \end{cases} \tag{3.55}$$

$$J^{-1}(m,n) = \begin{cases} \| Y(x) \| \sqrt{\beta(n)} & m = n, \\ 0 & m \neq n. \end{cases} \tag{3.56}$$

From the definition of $e_n(x)$ (3.32) it follows that

$$(e_k(x))_{\infty}^{k=1} = J(W^{-1})^T \, (Y(kx))_{\infty}^{k=1} . \tag{3.57}$$

Accordingly we have

$$(Y(kx))_{\infty}^{k=1} = W^T J^{-1} \, (e_k(x))_{\infty}^{k=1} . \tag{3.58}$$

The elements of the matrices $J(W^{-1})^T$ and $W^T J^{-1}$ are

$$\begin{aligned} J(W^{-1})^T \, (m,n) &= J(m,m)(W^{-1})^T(m,n) \\ &= \begin{cases} \frac{1}{\|Y(x)\|} \frac{1}{\sqrt{\beta(m)}} \mu(\frac{m}{n}) B(\frac{m}{n}) & n \mid m, \\ 0 & n \nmid m, \end{cases} \end{aligned}$$

$$\begin{aligned} W^T J^{-1} \, (m,n) &= W^T(m,n) J^{-1}(n,n) \\ &= \begin{cases} \| \, Y(x) \, \| \, \sqrt{\beta(n)} B(\frac{m}{n}) & n \mid m, \\ 0 & n \nmid m. \end{cases} \end{aligned}$$

Formula 3.4

$$(e_k(x))_{\infty}^{k=1} \; \underset{R^{-1}}{\overset{R}{\rightleftarrows}} \; (\sin kx)_{\infty}^{k=1} . \tag{3.59}$$

From Formula 3.1 and Formula 3.3, it follows that

$$R = J(W^{-1})^T W, \tag{3.60}$$

$$R^{-1} = W^{-1} W^T J^{-1}. \tag{3.61}$$

Their elements are

$$\begin{aligned} R(m,n) &= (J(W^{-1})^T W)(m,n) \\ &= \sum_{l=1}^{\infty} (J(W^{-1})^T)(m,l) W(l,n) \\ &= \frac{1}{\| \, Y(x) \, \|} \frac{1}{\sqrt{\beta(m)}} \sum_{l=1}^{\infty} W^{-1}(l,m) W(l,n) \\ &= \frac{1}{\| \, Y(x) \, \|} \frac{1}{\sqrt{\beta(m)}} \sum_{l \mid (m,n)} \mu(\frac{m}{l}) B(\frac{m}{l}) B(\frac{n}{l}), \end{aligned} \tag{3.62}$$

$$\begin{aligned} R^{-1}(m,n) &= W^{-1}W^{T}J^{-1}(m,n) \\ &= \sum_{l=1}^{\infty} W^{-1}(m,l)W^{T}(l,n)J^{-1}(n,n) \\ &= \| Y(x) \| \sqrt{\beta(n)} \sum_{m|l,n|l} \mu(\frac{l}{m})B(\frac{l}{m})B(\frac{l}{n}). \end{aligned} \tag{3.63}$$

On the other hand, from

$$\sin mx = \sum_{n=1}^{\infty} R^{-1}(m,n)e_n(x) \tag{3.64}$$

it follows that

$$\begin{aligned} R^{-1}(m,n) &= \langle \sin mx, e_n(x) \rangle \\ &= \frac{1}{\| Y(x) \|} \frac{1}{\sqrt{\beta(n)}} \sum_{d|n} \mu(\frac{n}{d})B(\frac{n}{d})\langle \sin mx, Y(dx) \rangle \\ &= \pi \frac{1}{\| Y(x) \|} \frac{1}{\sqrt{\beta(n)}} \sum_{d|(n,m)} \mu(\frac{n}{d})B(\frac{n}{d})B(\frac{m}{d}) \\ &= \pi R(n,m) = \pi R^{T}(m,n). \end{aligned}$$

Therefore we obtain the matrix identity

$$R^{-1} = \pi R^{T}, \tag{3.65}$$

or

$$R = \frac{1}{\pi}(R^{-1})^{T}. \tag{3.66}$$

The reason for this is that the sine function basis $\{\sin nx\}$ and the orthonormalized basis $\{e_n(x)\}$ are both orthogonal bases.

Finally we have

$$R^{-1} = W^{-1}W^{T}J^{-1} = \pi W^{T}W^{-1}J, \tag{3.67}$$

$$R = J(W^{-1})^{T}W = \frac{1}{\pi}J^{-1}W(W^{-1})^{T}. \tag{3.68}$$

In the next formula we shall use the above matrix identity.

Formula 3.5

$$(e_k(x))_\infty^{k=1} \underset{W^{-1}J}{\overset{J^{-1}W}{\rightleftarrows}} (g_n(x))_\infty^{k=1}. \tag{3.69}$$

By Formula 3.2 and Formula 3.4 we have

$$(e_k(x))_\infty^{k=1} = R(\sin kx)_\infty^{k=1} = R\pi W^T (g_k(x))_\infty^{k=1} = J^{-1}W(g_k(x))_\infty^{k=1},$$

and therefore

$$(g_k(x))_\infty^{k=1} = W^{-1}J(e_k(x))_\infty^{k=1}. \tag{3.70}$$

The elements of the matrices $J^{-1}W$ and $W^{-1}J$ are

$$J^{-1}W\ (m,n) = J^{-1}(m,m)W(m,n), \tag{3.71}$$

and

$$W^{-1}J\ (m,n) = W^{-1}(m,n)J(n,n). \tag{3.72}$$

Formula 3.6

$$(Y(kx))_\infty^{k=1} \underset{G^{-1}}{\overset{G}{\rightleftarrows}} (g_k(x))_\infty^{k=1}, \tag{3.73}$$

From Formulas 3.1,3.2 and Formulas 3.3,3.5 it follows that

$$G = \pi W W^T = W^T J^{-1} J^{-1} W, \tag{3.74}$$

$$G^{-1} = \frac{1}{\pi}(W^{-1})^T W^{-1} = W^{-1}JJ(W^{-1})^T. \tag{3.75}$$

The elements of the matrices G and G^{-1} are

$$\begin{aligned} G(m,n) &= \pi W W^T\ (m,n) \\ &= \pi \sum_{l=1}^{\infty} W(m,l)W^T(l,n) \\ &= \pi \sum_{m|k,n|k} B\left(\frac{k}{m}\right)B\left(\frac{k}{n}\right) \\ &= \|\, Y(x) \,\|^2\, B\left(\frac{mn}{(m,n)^2}\right), \end{aligned}$$

$$\begin{aligned}
G^{-1}(m,n) &= \frac{1}{\pi}(W^{-1})^T W^{-1}\ (m,n) \\
&= \frac{1}{\pi}\sum_{l=1}^{\infty}(W^{-1})^T(m,l)W^{-1}(l,n) \\
&= \frac{1}{\pi}\sum_{d|(m,n)}\mu(\frac{m}{d})\mu(\frac{n}{d})B(\frac{m}{d})B(\frac{n}{d}).
\end{aligned}$$

Summarily we can explicitly write the relations of the four bases associated with the WASCMFC function $Y(x) \in Y$ as the following (Fig.3.2):

1.
$(Y(kx))_{\infty}^{k=1} = W(\sin kx)_{\infty}^{k=1} = G(g_k(x))_{\infty}^{k=1} = W^T J^{-1}(e_k(x))_{\infty}^{k=1}$
or,
$Y(mx) = \sum_{l=1}^{\infty} B(l)\sin lmx = \sum_{n=1}^{\infty} G(m,n)g_n(x)$
$= \| Y(x) \| \sum_{d|m} B(\frac{m}{d})\sqrt{\beta(d)}e_d(x)$

2.
$(\sin kx)_{\infty}^{k=1} = \pi W^T(g_k(x))_{\infty}^{k=1} = R^{-1}(e_k(x))_{\infty}^{k=1} = W^{-1}(Y(kx))_{\infty}^{k=1}$
or,
$\sin mx = \pi\sum_{d|m} B(\frac{m}{d})g_d(x) = \sum_{n=1}^{\infty} R^{-1}(m,n)e_n(x)$
$= \sum_{l=1}^{\infty} \mu(l)B(l)Y(lmx)$

3.
$(g_k(x))_{\infty}^{k=1} = W^{-1}J(e_k(x))_{\infty}^{k=1} = G^{-1}(Y(kx))_{\infty}^{k=1} = \frac{1}{\pi}(W^{-1})^T(\sin kx)_{\infty}^{k=1}$
or,
$g_m(x) = \frac{1}{\|Y(x)\|}\sum_{l=1}^{\infty}\mu(l)B(l)\frac{1}{\sqrt{\beta(ml)}}e_{ml}(x) = \sum_{n=1}^{\infty} G^{-1}(m,n)Y(nx)$
$= \frac{1}{\pi}\sum_{d|m}\mu(\frac{m}{d})B(\frac{m}{d})\sin dx$

4.
$(e_k(x))_{\infty}^{k=1} = J(W^{-1})^T(Y(kx))_{\infty}^{k=1} = R(\sin kx)_{\infty}^{k=1} = J^{-1}W(g_k(x))_{\infty}^{k=1}$
or,
$e_m(x) = \frac{1}{\|Y(x)\|}\frac{1}{\sqrt{\beta(m)}}\sum_{d|m}\mu(\frac{m}{d})B(\frac{m}{d})Y(dx) = \sum_{n=1}^{\infty} R(m,n)\sin nx$
$= \| Y(x) \| \sqrt{\beta(m)}\sum_{l=1}^{\infty} B(l)g_{lm}(x).$

$(m = 1, 2, 3, \cdots.)$

Suppose that a given odd function $f(x) \in L^2[-\pi, \pi]$ is expressed as

$$\begin{aligned} f(x) &= D(1)Y(x) + D(2)Y(2x) + \cdots + D(n)Y(nx) + \cdots \\ &= t(1)g_1(x) + t(2)g_2(x) + \cdots + t(n)g_n(x) + \cdots \\ &= q(1)e_1(x) + q(2)e_2(x) + \cdots + q(n)e_n(x) + \cdots \\ &= b(1)\sin(x) + b(2)\sin(2x) + \cdots + b(n)\sin(nx) + \cdots . \end{aligned}$$

It follows easily that

$$\| f(x) \|^2 = \pi \sum_{n=1}^{\infty} b^2(n) = \sum_{n=1}^{\infty} q^2(n) = \sum_{n=1}^{\infty} D(n)t(n). \tag{3.76}$$

By relations between basis transforms and coordinate transforms, we can easily obtain the relations of the coordinates of $f(x)$ with respect to the four bases. See Figure 3.3.

In Figure 3.3, as an example, the coordinates with respect to the WASCMFC function basis, i.e., $\{D(n)\}$, and the coordinates with respect to the sine function basis, i.e., $\{b(n)\}$, are related by

$$(D(k))_{\infty}^{k=1} \underset{W^T}{\overset{(W^{-1})^T}{\rightleftarrows}} (b(k))_{\infty}^{k=1}. \tag{3.77}$$

As before, (3.77) means that

$$(D(k))_{\infty}^{k=1} = (W^{-1})^T (b(k))_{\infty}^{k=1}, \tag{3.78}$$

$$(b(k))_{\infty}^{k=1} = W^T (D(k))_{\infty}^{k=1}. \tag{3.79}$$

We can rewrite the relations explicitly as

$$D(m) = \sum_{d|m} \mu(\frac{m}{d}) B(\frac{m}{d}) b(d), \tag{3.80}$$

$$b(m) = \sum_{d|m} B(\frac{m}{d}) D(d), \tag{3.81}$$

for $m = 1, 2, 3, \cdots$, or simply

$$D = B^{-1} * b, \tag{3.82}$$

$$b = B * D. \tag{3.83}$$

2. Now we suppose that $Y_{\rm I}(x)$ and $Y_{\rm II}(x)$ are two odd WASCMFC functions, i.e. $Y_{\rm I}(x), Y_{\rm II}(x) \in Y$. Let $B_{\rm I}(n)$ and $B_{\rm II}(n)$ be the Fourier coefficients of the two functions, respectively.

There are seven bases in the odd function subspace, i.e.

$$\{Y_{\rm I}(nx)\},\ \{g_{\rm In}(x)\},\ \{e_{\rm In}(x)\},\ \{Y_{\rm II}(nx)\},\ \{g_{\rm IIn}(x)\},\ \{e_{\rm IIn}(x)\},\ \{\sin nx\}, \tag{3.84}$$

which are related via the sine function basis $\{\sin nx\}$, as shown in Figure 3.4.

As an example, let us consider the relation between the WASCMFC function bases $\{Y_{\rm I}(nx)\}$ and $\{Y_{\rm II}(nx)\}$. From Figure 3.4, we obtain that

$$(Y_{\rm I}(kx))_{\infty}^{k=1} \underset{W_{\rm II}W_{\rm I}^{-1}}{\overset{W_{\rm I}W_{\rm II}^{-1}}{\rightleftarrows}} (Y_{\rm II}(kx))_{\infty}^{k=1}. \tag{3.85}$$

This means that

$$(Y_{\rm I}(kx))_{\infty}^{k=1} = W_{\rm I}W_{\rm II}^{-1}(Y_{\rm II}(kx))_{\infty}^{k=1}, \tag{3.86}$$

$$(Y_{\rm II}(kx))_{\infty}^{k=1} = W_{\rm II}W_{\rm I}^{-1}(Y_{\rm I}(kx))_{\infty}^{k=1}. \tag{3.87}$$

In Dirichlet multiplication, we write (3.86) and (3.87) as

$$Y_{\rm I}(mx) = \sum_{l=1}^{\infty} B_{\rm I} * B_{\rm II}^{-1}\ (l)\ Y_{\rm II}(lmx), \tag{3.88}$$

$$Y_{\rm II}(mx) = \sum_{l=1}^{\infty} B_{\rm II} * B_{\rm I}^{-1}\ (l)\ Y_{\rm I}(lmx), \tag{3.89}$$

where $m = 1, 2, 3, \cdots$.

Example. The triangular wave $Y_{\rm tri}(x)$ and the II-trapezoidal wave $Y_{\rm IItra}(x)$ are related by

$$Y_{\rm tri}(x) = \sum_{l=1}^{\infty} B_{\rm tri} * B_{\rm IItra}^{-1}\ (l)\ Y_{\rm IItra}(lx), \tag{3.90}$$

$$Y_{\rm IItra}(x) = \sum_{l=1}^{\infty} B_{\rm IItra} * B_{\rm tri}^{-1}\ (l)\ Y_{\rm tri}(lx). \tag{3.91}$$

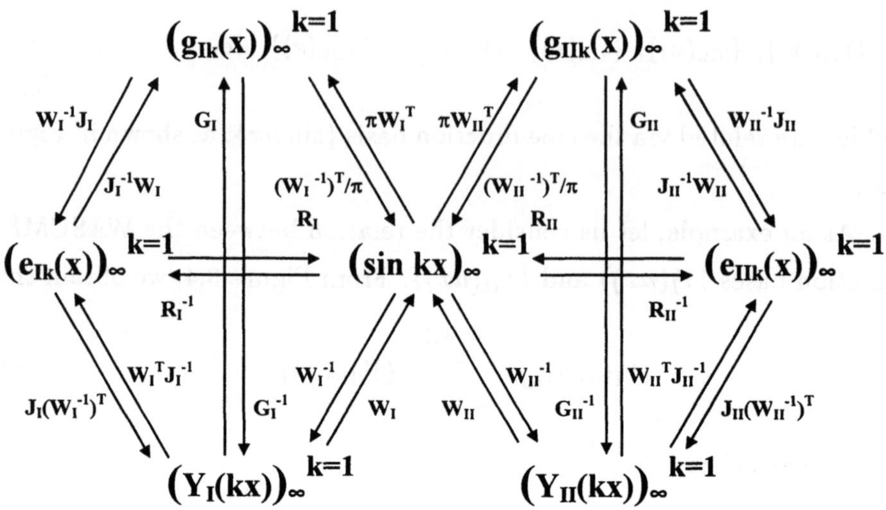

Figure 3.4: The relations of the seven bases.

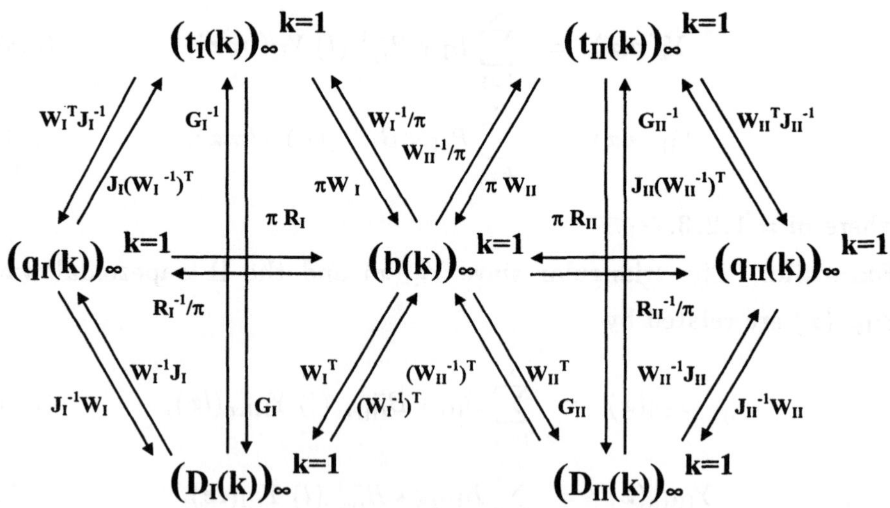

Figure 3.5: The relations of the seven coordinates.

Suppose that $f(x)$ is an odd function in $L^2[-\pi,\pi]$. Let its coordinates with respect to the seven bases (3.84) be

$$\{D_{\rm I}(n)\},\quad \{t_{\rm I}(n)\},\quad \{q_{\rm I}(n)\},\quad \{D_{\rm II}(n)\},\quad \{t_{\rm II}(n)\},\quad \{q_{\rm II}(n)\},\quad \{b(n)\},$$

respectively.

These coordinates are related via the Fourier coefficients $\{b(n)\}$, as shown in Figure 3.5.

As an example, let us consider the relations between $\{D_{\rm I}(n)\}$, and $\{D_{\rm II}(n)\}$. From Figure 3.5, we obtain that $\{D_{\rm I}(n)\}$, and $\{D_{\rm II}(n)\}$ are related by

$$(D_{\rm I}(k))_{\infty}^{k=1} \underset{(W_{\rm II}^{-1})^T W_{\rm I}^T}{\overset{(W_{\rm I}^{-1})^T W_{\rm II}^T}{\rightleftarrows}} (D_{\rm II}(k))_{\infty}^{k=1}. \tag{3.92}$$

This means that

$$(D_{\rm I}(k))_{\infty}^{k=1} = (W_{\rm I}^{-1})^T W_{\rm II}^T (D_{\rm II}(k))_{\infty}^{k=1}, \tag{3.93}$$
$$(D_{\rm II}(k))_{\infty}^{k=1} = (W_{\rm II}^{-1})^T W_{\rm I}^T (D_{\rm I}(k))_{\infty}^{k=1}. \tag{3.94}$$

In Dirichlet multiplication we write (3.93) and (3.94) as

$$D_{\rm I} = B_{\rm I}^{-1} * B_{\rm II} * D_{\rm II}, \tag{3.95}$$
$$D_{\rm II} = B_{\rm II}^{-1} * B_{\rm I} * D_{\rm I}. \tag{3.96}$$

We consider the basis transforms and the coordinate transforms only in the odd function subspace of $L^2[-\pi,\pi]$, because the results in the even function subspace can be obtained similarly.

3.4 Discrete Triangular Wave Transform and Trapezoidal Wave Transform

In this section we shall generalize the discrete Fourier transform (DFT) to the discrete triangular wave transform and trapezoidal wave transform.

R^N denotes the N-dimensional Euclidean space. Suppose $f(x)$ is a function with period 2π, and it is known only at the N points $x_k = \frac{k}{N}2\pi$, where $k = 0, 1, 2, \cdots, N-1$. For convenience we introduce the notation

$$(f(x))_N = \begin{pmatrix} f(x_0) \\ f(x_1) \\ \vdots \\ f(x_{N-1}) \end{pmatrix}, \qquad (3.97)$$

which denotes a column vector consisting of N values of $f(x)$. The inner product in R^N is denoted by

$$\langle f(x), g(x) \rangle_N = \sum_{k=0}^{N-1} f(x_k) g(x_k). \qquad (3.98)$$

If two periodic functions $f(x)$ and $g(x)$ are equal at all the N points, i.e.,

$$f(x_k) = g(x_k) \quad \text{for} \quad k = 0, 1, 2, \cdots, N-1, \qquad (3.99)$$

we write

$$(f(x))_N = (g(x))_N, \qquad (3.100)$$

or

$$f(x) \overset{N}{=} g(x). \qquad (3.101)$$

I. The case $N = 2m + 1$.

Our question is: if a periodic function $f(x)$ with the period 2π is known at the $N = 2m + 1$ points, for given functions $X(x) \in X$ and $Y(x) \in Y$, how to determine the coefficients $C(0), C(n), D(n)$ such that

$$f(x) \overset{N}{=} C(0) + \sum_{n=1}^{m} C(n)X(nx) + D(n)Y(nx) \ ? \qquad (3.102)$$

In case of $X(x) = \cos x$, $Y(x) = \sin x$, the answer is clear. According to discrete Fourier transform (DFT) [3], we have

$$(f(x))_N = a(0)(1)_N + \sum_{n=1}^{m} a(n)(\cos nx)_N + b(n)(\sin nx)_N, \qquad (3.103)$$

where

$$
\begin{aligned}
a(0) &= \frac{1}{N}\langle 1, f(x)\rangle_N = \frac{1}{N}\sum_{k=0}^{N-1} f(x_k), \\
a(n) &= \frac{2}{N}\langle \cos nx, f(x)\rangle_N = \frac{2}{N}\sum_{k=0}^{N-1} f(x_k)\cos nx_k, \\
b(n) &= \frac{2}{N}\langle \sin nx, f(x)\rangle_N = \frac{2}{N}\sum_{k=0}^{N-1} f(x_k)\sin nx_k,
\end{aligned}
$$

because the vector system

$$(1)_N, (\cos x)_N, \cdots, (\cos mx)_N, (\sin x)_N, \cdots, (\sin mx)_N \tag{3.104}$$

is an orthogonal basis of R^N.

Next let us consider the general case. The Fourier series of $X(x) \in X$ and $Y(x) \in Y$ are

$$X(x) = \sum_{n=1}^{\infty} A(n)\cos nx, \tag{3.105}$$

$$Y(x) = \sum_{n=1}^{\infty} B(n)\sin nx. \tag{3.106}$$

Since

$$(\cos(n+N)x)_N = (\cos nx)_N, \tag{3.107}$$

$$(\sin(n+N)x)_N = (\sin nx)_N, \tag{3.108}$$

it follows that

$$(X(ix))_N = \sum_{n=0}^{m} A_i(n)(\cos nx)_N, \tag{3.109}$$

$$(Y(ix))_N = \sum_{n=1}^{m} B_i(n)(\sin nx)_N, \tag{3.110}$$

where

$$A_i(0) = \sum_{il\equiv 0} A(l)$$

$$\begin{aligned} A_i(n) &= \sum_{il\equiv n} A(l) + \sum_{il\equiv -n} A(l) \\ B_i(n) &= \sum_{il\equiv n} B(l) - \sum_{il\equiv -n} B(l) \end{aligned}$$

with $1 \le i \le m$. $il \equiv n$ means the residue of il to modulus N is n.

Similarly, from (3.14) and (3.15) it follows that

$$(\cos ix)_N = \sum_{n=0}^{m} A_i^{-1}(n)(X(nx))_N, \tag{3.111}$$

$$(\sin ix)_N = \sum_{n=1}^{m} B_i^{-1}(n)(Y(nx))_N, \tag{3.112}$$

where

$$\begin{aligned} A_i^{-1}(0) &= \sum_{il\equiv 0} A^{-1}(l) = \sum_{il\equiv 0} \mu(l)A(l) \\ A_i^{-1}(n) &= \sum_{il\equiv n} A^{-1}(l) + \sum_{il\equiv -n} A^{-1}(l) = \sum_{il\equiv n} \mu(l)A(l) + \sum_{il\equiv -n} \mu(l)A(l) \\ B_i^{-1}(n) &= \sum_{il\equiv n} B^{-1}(l) - \sum_{il\equiv -n} B^{-1}(l) = \sum_{il\equiv n} \mu(l)B(l) - \sum_{il\equiv -n} \mu(l)B(l) \end{aligned}$$

with $1 \le i \le m$.

Since (3.104) is an orthogonal basis, it follows from (3.111) and (3.112) that the vector system

$$(1)_N, (X(x))_N, \cdots, (X(mx))_N, (Y(x))_N, \cdots, (Y(mx))_N \tag{3.113}$$

is linearly independent and therefore it is another basis of R^N.

In matrix notation, the two bases (3.104) and (3.113) are related by

$$\left((1)_N, (X(x))_N, \cdots, (X(mx))_N, (Y(x))_N, \cdots, (Y(mx))_N \right)$$

$$= \left((1)_N, (\cos x)_N, \cdots, (\cos mx)_N, (\sin x)_N, \cdots, (\sin mx)_N \right) P,$$

and

$$\begin{pmatrix} (1)_N, (\cos x)_N, \cdots, (\cos mx)_N, (\sin x)_N, \cdots, (\sin mx)_N \end{pmatrix}$$
$$= \begin{pmatrix} (1)_N, (X(x))_N, \cdots, (X(mx))_N, (Y(x))_N, \cdots, (Y(mx))_N \end{pmatrix} P^{-1},$$

where

$$P = \left(\begin{array}{cccc|ccc} 1 & A_1(0) & \cdots & A_m(0) & & & \\ 0 & A_1(1) & \cdots & A_m(1) & & 0 & \\ \vdots & \vdots & \vdots & \vdots & & & \\ 0 & A_1(m) & \cdots & A_m(m) & & & \\ & & & & B_1(1) & \cdots & B_m(1) \\ & & 0 & & \vdots & \vdots & \vdots \\ & & & & B_1(m) & \cdots & B_m(m) \end{array} \right),$$

$$P^{-1} = \left(\begin{array}{cccc|ccc} 1 & A_1^{-1}(0)X(0) & \cdots & A_m^{-1}(0)X(0) & & & \\ 0 & A_1^{-1}(1) & \cdots & A_m^{-1}(1) & & 0 & \\ \vdots & \vdots & \vdots & \vdots & & & \\ 0 & A_1^{-1}(m) & \cdots & A_m^{-1}(m) & & & \\ & & & & B_1^{-1}(1) & \cdots & B_m^{-1}(1) \\ & & 0 & & \vdots & \vdots & \vdots \\ & & & & B_1^{-1}(m) & \cdots & B_m^{-1}(1) \end{array} \right).$$

Equation (3.102) can be rewritten as

$$(f(x))_N = C(0)(1)_N + \sum_{n=1}^{m} C(n)(X(nx))_N + D(n)(Y(nx))_N, \quad (3.114)$$

and the two groups of coefficients are related by

$$\begin{pmatrix} C(0) \\ C(1) \\ \vdots \\ C(m) \\ D(1) \\ \vdots \\ D(m) \end{pmatrix} = P^{-1} \begin{pmatrix} a(0) \\ a(1) \\ \vdots \\ a(m) \\ b(1) \\ \vdots \\ b(m) \end{pmatrix}, \tag{3.115}$$

$$\begin{pmatrix} a(0) \\ a(1) \\ \vdots \\ a(m) \\ b(1) \\ \vdots \\ b(m) \end{pmatrix} = P \begin{pmatrix} C(0) \\ C(1) \\ \vdots \\ C(m) \\ D(1) \\ \vdots \\ D(m) \end{pmatrix}. \tag{3.116}$$

If we define

$$\begin{aligned} u_0(x) &= \frac{1}{N} + \frac{2}{N} X(0) \sum_{n=1}^{m} A_n^{-1}(0) \cos nx, \\ u_i(x) &= \frac{2}{N} \sum_{n=1}^{m} A_n^{-1}(i) \cos nx, \\ v_i(x) &= \frac{2}{N} \sum_{n=1}^{m} B_n^{-1}(i) \sin nx, \end{aligned} \tag{3.117}$$

then the coefficients in (3.102) or (3.114) can be uniquely determined by

$$\begin{aligned} C(0) &= \langle u_0(x), f(x) \rangle_N = \sum_{k=0}^{N-1} u_0(x_k) f(x_k), \\ C(i) &= \langle u_i(x), f(x) \rangle_N = \sum_{k=0}^{N-1} u_i(x_k) f(x_k), \\ D(i) &= \langle v_i(x), f(x) \rangle_N = \sum_{k=0}^{N-1} v_i(x_k) f(x_k), \end{aligned} \tag{3.118}$$

where $i = 1, 2, \cdots, m$.

II. The case $N = 2m$

Our question is: if a periodic function $f(x)$ with the period 2π is known at the $N = 2m$ points, for given functions $X(x) \in X$ and $Y(x) \in Y$, how to determine the coefficients $C(0), C(n), D(n)$ such that

$$f(x) \stackrel{N}{=} C(0) + \sum_{n=1}^{m-1} C(n)X(nx) + D(n)Y(nx) + C(m)X(mx) \quad ? \quad (3.119)$$

The result in this case is similar. We write

$$\begin{aligned}(f(x))_N &= a(0)(1)_N + \sum_{n=1}^{m-1} a(n)(\cos nx)_N + b(n)(\sin nx)_N \\ &\quad + a(m)(\cos mx)_N \\ &= C(0)(1)_N + \sum_{n=1}^{m-1} C(n)(X(nx))_N + D(n)(Y(nx))_N \\ &\quad + C(m)(X(mx))_N,\end{aligned}$$

where [3]

$$\begin{aligned} a(0) &= \frac{1}{N}\langle 1, f(x)\rangle_N = \frac{1}{N}\sum_{k=0}^{N-1} f(x_k), \\ a(n) &= \frac{2}{N}\langle \cos nx, f(x)\rangle_N = \frac{2}{N}\sum_{k=0}^{N-1} f(x_k)\cos nx_k, \\ a(m) &= \frac{1}{N}\langle \cos mx, f(x)\rangle_N = \frac{1}{N}\sum_{k=0}^{N-1} f(x_k)\cos mx_k, \\ b(n) &= \frac{2}{N}\langle \sin nx, f(x)\rangle_N = \frac{2}{N}\sum_{k=0}^{N-1} f(x_k)\sin nx_k. \end{aligned}$$

From (3.14) and (3.15), it follows that

$$\begin{aligned} (\cos ix)_N &= A_i^{-1}(0)X(0)(1)_N + \sum_{n=1}^{m-1} A_i^{-1}(n)(X(nx))_N \\ &\quad + A_i^{-1}(m)(X(mx))_N, \\ (\sin ix)_N &= \sum_{n=1}^{m-1} B_i^{-1}(n)(Y(nx))_N, \end{aligned}$$

where

$$\begin{aligned}
A_i^{-1}(0) &= \sum_{il\equiv 0} A^{-1}(l), && (3.120)\\
A_i^{-1}(n) &= \sum_{il\equiv n} A^{-1}(l) + \sum_{il\equiv -n} A^{-1}(l), && (3.121)\\
A_i^{-1}(m) &= \sum_{il\equiv m} A^{-1}(l), && (3.122)\\
B_i^{-1}(n) &= \sum_{il\equiv n} B^{-1}(l) - \sum_{il\equiv -n} B^{-1}(l). && (3.123)
\end{aligned}$$

In matrix notation, we have

$$\begin{aligned}
&\left((1)_N, (\cos x)_N, \cdots, (\cos mx)_N, (\sin x)_N, \cdots, (\sin(m-1)x)_N \right)\\
&= \left((1)_N, (X(x))_N, \cdots, (X(mx))_N, (Y(x))_N, \cdots, (Y((m-1)x))_N \right) Q,
\end{aligned}$$

$$\begin{pmatrix} C(0) \\ C(1) \\ \vdots \\ C(m) \\ D(1) \\ \vdots \\ D(m-1) \end{pmatrix} = Q \begin{pmatrix} a(0) \\ a(1) \\ \vdots \\ a(m) \\ b(1) \\ \vdots \\ b(m-1) \end{pmatrix},$$

where

$$Q = \begin{pmatrix}
1 & A_1^{-1}(0)X(0) & \cdots & A_m^{-1}(0)X(0) & & & \\
0 & A_1^{-1}(1) & \cdots & A_m^{-1}(1) & & 0 & \\
\vdots & \vdots & \vdots & \vdots & & & \\
0 & A_1^{-1}(m) & \cdots & A_m^{-1}(m) & & & \\
 & & & & B_1^{-1}(1) & \cdots & B_{m-1}^{-1}(1) \\
 & & 0 & & \vdots & \vdots & \vdots \\
 & & & & B_1^{-1}(m-1) & \cdots & B_{m-1}^{-1}(1)
\end{pmatrix}.$$

If we define

$$\begin{aligned}
u_0(x) &= \frac{1}{N} + \frac{2}{N} X(0) \sum_{i=1}^{m-1} A_i^{-1}(0) \cos ix + \frac{1}{N} X(0) A_m^{-1}(0) \cos mx, \\
u_n(x) &= \frac{2}{N} \sum_{i=1}^{m-1} A_i^{-1}(n) \cos ix + \frac{1}{N} A_m^{-1}(0) \cos mx, \quad (n = 1, 2, ..., m) \\
v_n(x) &= \frac{2}{N} \sum_{i=1}^{m-1} B_i^{-1}(n) \sin ix, \quad (n = 1, 2, ... m-1)
\end{aligned}$$

then the coefficients in (3.119) can be determined by

$$\begin{aligned}
C(n) &= \langle u_n(x), f(x) \rangle_N = \sum_{k=0}^{N-1} u_n(x_k) f(x_k), \quad (n = 0, 1, ..., m) \\
D(n) &= \langle v_n(x), f(x) \rangle_N = \sum_{k=0}^{N-1} v_n(x_k) f(x_k). \quad (n = 1, 2 ..., m-1)
\end{aligned}$$

SUMMARY

In this chapter, we prove that triangular wave system and trapezoidal wave system are unconditional basis of $L^2[-\pi, \pi]$ and present the relation between triangular wave series, trapezoidal wave series and Fourier series. As a generalization of discrete Fourier transform, discrete triangular wave transform and trapezoidal wave transform are discussed.

Unsolved Problem:

1. Does the triangular wave series or trapezoidal wave series of $f(x) \in L^2[-\pi, \pi]$ (3.22) converges to $f(x)$ almost everywhere? This is a general version of Luzin problem.

Chapter 4

FREQUENCY ANALYSIS BASED ON GENERAL PERIODIC FUNCTIONDS

Electrical waveforms may have various shapes, and therefore it is necessary to study the frequency analysis based on general periodic functions. In this chapter, we introduce the frequency series and frequency transform based on general periodic functions. Considering the fact that the theory for Fourier series in $L^2[-\pi, \pi]$ and the theory for Fourier transform in $L^2(-\infty, +\infty)$ are both simple and elegant, we shall develop a theory for general frequency series in $L^2[-\pi, \pi]$ and a theory for general frequency transform in $L^2(-\infty, +\infty)$. In the following, we shall discuss when a frequency system is a complete system or an unconditional basis in $L^2[-\pi, \pi]$, and when a frequency transform can be carried out in $L^2(-\infty, +\infty)$. We present a number of complete systems and unconditional bases formed by common waveforms. Almost every common waveform is considered as an example

It is well-known that $L^2[-\pi, +\pi]$ can be decomposed into three mutually orthogonal subspaces:

$$\begin{aligned} L^2[-\pi,+\pi] &= \text{span}\{1\} \oplus L^2_{\text{even}}[-\pi,+\pi] \oplus L^2_{\text{odd}}[-\pi,+\pi], \\ &= \text{span}\{1\} \oplus \overline{\text{span}\{\cos nx\}_{n=1}^{\infty}} \oplus \overline{\text{span}\{\sin nx\}_{n=1}^{\infty}}, \end{aligned}$$

where span{1} is the constant function subspace, $L^2_{\text{even}}[-\pi,+\pi]$ is the subspace of even functions with the mean-value 0, and $L^2_{\text{odd}}[-\pi,+\pi]$ is the subspace of odd functions.

4.1 A Frequency System in $L^2[-\pi,\pi]$

In this section we suppose that the function

$$F(x) = \sum_{n=1}^{\infty} A(n)\cos(nx) + B(n)\sin(nx), \tag{4.1}$$

is a given nonzero element in $L^2[-\pi,\pi]$, where $A(n)$ and $B(n)$ denote its Fourier coefficients.

We shall show that the frequency system based on the function $F(x)$,

$$F(x), F(2x), F(3x), \cdots, F(nx), \cdots \tag{4.2}$$

is linearly independent, weakly convergent to zero, and incomplete in $L^2[-\pi,\pi]$.

Proposition 4.1 We have

$$\sum_{n=1}^{\infty} c(n)F(nx) = 0 \Longleftrightarrow c = 0, \tag{4.3}$$

where $c(n)$ is a real-valued arithmetical function.

Proof. If

$$c = 0, \tag{4.4}$$

the conclusion is trivially true.

Let us consider the case

$$\sum_{n=1}^{\infty} c(n)F(nx) = 0. \tag{4.5}$$

Putting (4.1) in (4.5), we have

$$\sum_{n=1}^{\infty}(a(n)\cos nx + b(n)\sin nx) = 0, \tag{4.6}$$

where

$$\begin{aligned} a(n) &= \sum_{d|n} A(\frac{n}{d})c(d) &= A * c\ (n), \\ b(n) &= \sum_{d|n} B(\frac{n}{d})c(d) &= B * c\ (n), \end{aligned}$$

From (4.6), we obtain that

$$a = b = 0.$$

Since $F(x)$ is a nonzero element, we have

$$A \neq 0, \quad \text{or } B \neq 0. \tag{4.7}$$

By Proposition 1.3, we obtain that

$$c = 0, \tag{4.8}$$

This completes the proof. □

Consequence. If a function $\psi(x) \in L^2[-\pi, \pi]$ can be expressed as a frequency series based on F(x):

$$\psi(x) = c(1)F(x) + c(2)F(2x) + c(3)F(3x) + \cdots + c(n)F(nx) + \cdots, \tag{4.9}$$

then the coefficients $c(n)$ are unique.

Proposition 4.2 The frequency system $\{F(nx)\}_{n=1}^{\infty}$ converges weakly to 0.

Proof. Suppose that $\psi(x) \in L^2[-\pi, +\pi]$ is an arbitrary function, and its Fourier series is

$$\psi(x) = a_0 + \sum_{l=1}^{\infty} a(l)\cos(lx) + b(l)\sin(lx). \tag{4.10}$$

We have

$$\begin{aligned}
& \mid \langle F(nx), \psi(x) \rangle \mid \\
= & \ \pi \mid \sum_{l=1}^{\infty} A(l) a(nl) + \sum_{l=1}^{\infty} B(l) b(nl) \mid \\
\leq & \ \pi \sqrt{\sum_{l=1}^{\infty} A^2(l)} \sqrt{\sum_{l=1}^{\infty} a^2(nl)} + \pi \sqrt{\sum_{l=1}^{\infty} B^2(l)} \sqrt{\sum_{l=1}^{\infty} b^2(nl)} \\
\leq & \ \pi \sqrt{\sum_{l=1}^{\infty} A^2(l)} \sqrt{\sum_{l=n}^{\infty} a^2(l)} + \pi \sqrt{\sum_{l=1}^{\infty} B^2(l)} \sqrt{\sum_{l=n}^{\infty} b^2(l)} \\
\to & \ 0 \quad (\text{ as } n \to \infty).
\end{aligned}$$

Therefore we have

$$\langle F(nx), \psi(x) \rangle \longrightarrow 0 \quad (\text{ as } n \to \infty). \tag{4.11}$$

This completes the proof. □

Proposition 4.3 The frequency system $\{F(nx)\}_{n=1}^{\infty}$ is incomplete in $L^2[-\pi, +\pi]$.

Proof.

There are two cases.

1. If $A(1) \neq 0$ or $B(1) \neq 0$, then there exists a nonzero element

$$\psi(x) = -B(1)\cos x + A(1)\sin x$$

such that

$$\int_{-\pi}^{\pi} F(nx)\psi(x)dx = 0 \ \text{ for n} = 1, 2, 3, \cdots. \tag{4.12}$$

2. If $A(1) = B(1) = 0$, we have

$$\int_{-\pi}^{\pi} F(nx)\cos x dx = 0 \ \text{ for n} = 1, 2, 3, \cdots. \tag{4.13}$$

In one word, the frequency system $\{F(nx)\}_{n=1}^{\infty}$ is incomplete in $L^2[-\pi, +\pi]$. □

Now that the frequency system is incomplete in $L^2[-\pi, +\pi]$, we shall develop the problem in its odd function subspace $L^2_{\text{odd}}[-\pi, +\pi]$ in the following section.

4.2 A Frequency System in $L^2_{\text{odd}}[-\pi, +\pi]$

In this section we suppose that the function

$$Y(x) = \sum_{n=1}^{\infty} B(n) \sin nx \tag{4.14}$$

is a given nonzero element in $L^2_{\text{odd}}[-\pi, +\pi]$, where $B(n)$ denotes its Fourier coefficients.

We shall show that the frequency system $\{Y(nx)\}_{n=1}^{\infty}$ may be an orthogonal basis, an unconditional basis, a complete system, or an incomplete system in $L^2_{\text{odd}}[-\pi, +\pi]$. Let us see a simple example.

Example.

If $Y(x) = \sin x + k \sin 2x$, where k is a real number, then in $L^2_{\text{odd}}[-\pi, +\pi]$ the frequency system $\{Y(nx)\}_{n=1}^{\infty}$ is:

(1) an orthogonal basis when $k = 0$;

(2) an unconditional basis when $0 <| k |< 1$;

(3) a complete system, but not a basis, when $| k |= 1$;

(4) an incomplete system when $| k |> 1$.

Proof of the example.

(1) This is clear.

(2) This follows from Proposition 4.18, which we shall introduce.

(3) Let us consider the completeness first.

Suppose that $\psi(x) \in L^2_{\text{odd}}[-\pi, +\pi]$ is a function such that

$$\int_{-\pi}^{\pi} \psi(x) Y(nx) dx = 0 \quad \text{for } n = 1, 2, 3, \cdots. \tag{4.15}$$

The Fourier series of $\psi(x)$ can be expressed as

$$\psi(x) = \sum_{n=1}^{\infty} b(n) \sin nx. \tag{4.16}$$

From the relation (4.15), we obtain that

$$b(n) + kb(2n) = 0, \tag{4.17}$$

or,

$$b(n) = (-k)b(2n), \tag{4.18}$$

for any positive integers n.

Thus for an arbitrary positive integer l, we have

$$b(l) = (-k)b(2l) = (-k)^2 b(2^2 l) = \cdots = (-k)^m b(2^m l) = \cdots. \tag{4.19}$$

Since $b(n) \to 0$ as $n \to \infty$, we have

$$b(l) = 0. \tag{4.20}$$

Since l is an arbitrary positive integer, we obtain $\psi(x) = 0$. This completes the proof of the completeness.

However, the frequency system is not a basis of $L^2_{\text{odd}}[-\pi, +\pi]$.

If it is a basis, the function $\sin x$ can be expressed as a frequency series:

$$\sin x = \sum_{n=1}^{\infty} D(n) Y(nx), \tag{4.21}$$

where $D(n)$ are coefficients.

Putting $Y(x) = \sin x + k \sin 2x$ in (4.21), we obtain that

$$\begin{aligned} \sin x &= D(1)\sin x + (kD(1) + D(2))\sin 2x + D(3)\sin 3x \\ &\quad + (kD(2) + D(4))\sin 4x + \cdots. \end{aligned}$$

Considering the coefficients of the terms $\sin 2^m x$ ($m = 0, 1, 2, \cdots$.), we obtain that

$$D(1) = 1,$$

and

$$D(2^m) + kD(2^{m-1}) = 0, \quad \text{for } m = 1, 2, 3, \cdots.$$

Thus we obtain that

$$D(2^m) = (-k)^m, \quad \text{for } m = 1, 2, 3, \cdots. \tag{4.22}$$

Therefore $D(2^m)$ does not converge to 0 as m increases, and hence $D(n)$ does not converge to 0 either as n increases. This is a contradiction.

So the conclusion should be that the frequency system is not a basis of $L^2_{\text{odd}}[-\pi, +\pi]$.

(4) The frequency system is not complete in $L^2_{\text{odd}}[-\pi, +\pi]$, because there exists a function

$$\begin{aligned} \psi(x) &= \sin x + (-\frac{1}{k}) \sin 2x + (-\frac{1}{k})^2 \sin 4x + (-\frac{1}{k})^3 \sin 8x + \cdots \\ &\quad + (-\frac{1}{k})^m \sin 2^m x + \cdots \end{aligned}$$

such that

$$\int_{-\pi}^{\pi} \psi(x) Y(nx) dx = 0 \quad \text{for } n = 1, 2, 3, \cdots. \tag{4.23}$$

This completes the proof of the example. □

Proposition 4.4 The frequency system $\{Y(nx)\}_{n=1}^{\infty}$ is an orthonormal basis of $L^2_{\text{odd}}[-\pi, +\pi]$ if and only if

$$Y(x) = \pm \frac{1}{\sqrt{\pi}} \sin x. \tag{4.24}$$

Proof. If part is clear.

Next let us consider the only-if part. If $\{Y(nx)\}_{n=1}^{\infty}$ is an orthonormal basis of $L^2_{\text{odd}}[-\pi, +\pi]$, then the function $\sin x$ has a frequency series based on $Y(x)$:

$$\sin x = \sum_{n=1}^{\infty} D(n) Y(nx), \tag{4.25}$$

where the coefficients D(n) are determined by

$$D(n) = \int_{-\pi}^{\pi} \sin x Y(nx) dx. \tag{4.26}$$

It follows that

$$\begin{aligned} D(1) &= \pi B(1), \\ D(n) &= 0 \ \text{ for } n = 2, 3, 4, \cdots. \end{aligned}$$

Thus we have

$$\begin{aligned} \sin x &= \pi B(1) Y(x) \\ &= \pi B(1) \sum_{n=1}^{\infty} B(n) \sin nx. \end{aligned} \tag{4.27}$$

From (4.27), we obtain that

$$\begin{aligned} B(1) &= \pm\frac{1}{\sqrt{\pi}}, \\ B(n) &= 0 \text{ for } n = 2, 3, 4, \cdots, \end{aligned} \tag{4.28}$$

i.e.,

$$Y(x) = \pm\frac{1}{\sqrt{\pi}} \sin x. \quad \square \tag{4.29}$$

Proposition 4.5 The frequency system $\{Y(nx)\}_{n=1}^{\infty}$ is incomplete in $L^2_{\text{odd}}[-\pi, +\pi]$ if $Y(x)$'s first Fourier coefficient $B(1) = 0$.

Proof. It is because

$$\int_{-\pi}^{\pi} Y(nx) \sin x dx = 0, \tag{4.30}$$

for $n = 1, 2, 3, \cdots$. $\square$

Proposition 4.6 If $B(1) \neq 0$, then the frequency system

$$Y(x), Y(2x), Y(3x), \cdots, Y(nx), \cdots \tag{4.31}$$

and the function system

$$g_1(x), g_2(x), g_3(x), \cdots, g_n(x), \cdots \tag{4.32}$$

are biorthogonal, where the functions $g_n(x)$ are defined by

$$g_n(x) = \frac{1}{\pi} \sum_{d|n} B^{-1}(\frac{n}{d}) \sin dx \tag{4.33}$$

for $n = 1, 2, 3, \cdots$.

Proof. For any two positive integers m and n, we have

$$\begin{aligned} & \langle Y(mx), g_n(x) \rangle \\ = & \langle \sum_{k=1}^{\infty} B(k) \sin(mkx), \frac{1}{\pi} \sum_{d|n} B^{-1}(\frac{n}{d}) \sin dx \rangle \\ = & \sum_{m|d, d|n} B^{-1}(\frac{n}{d}) B(\frac{d}{m}) \end{aligned}$$

$$
\begin{aligned}
&= \begin{cases} 0 & m \nmid n \\ 0 & \frac{n}{m} = 2, 3, \cdots \\ 1 & \frac{n}{m} = 1 \end{cases} \\
&= \begin{cases} 0 & m \neq n \\ 1 & m = n \end{cases} \\
&= \delta_{mn}.
\end{aligned}
$$

This completes the proof. □

Proposition 4.7 If a function $\psi(x) \in L^2_{\text{odd}}[-\pi, +\pi]$ has a frequency series based on the function $Y(x)$ (with $B(1) \neq 0$):

$$\psi(x) = \sum_{n=1}^{\infty} D(n) Y(nx), \tag{4.34}$$

then the coefficients $D(n)$ can be determined uniquely by

$$D(n) = \int_{-\pi}^{\pi} \psi(x) g_n(x) dx, \tag{4.35}$$

where $g_n(x)$ are the biorthogonal functions defined by (4.33).

Proof. This proposition follows from Proposition 4.6 obviously. □

In the following sections, we shall discuss when a frequency system is a complete system or an unconditional basis.

4.3 A Complete System in $L^2_{\text{odd}}[-\pi, +\pi]$

We shall discuss when a frequency system is a complete system in $L^2_{\text{odd}}[-\pi, +\pi]$.

Definition. The subset W of $L^2_{\text{odd}}[-\pi, +\pi]$ is defined by

$$
\begin{aligned}
W = \{Y(x) \in L^2_{\text{odd}}[-\pi, +\pi] : \ & Y(x) = \sum_{n=1}^{\infty} B(n) \sin nx, \text{ with} \\
& \sum_{l=1}^{\infty} \Big(\sum_{mn=l} | B^{-1}(n) B(m) | \Big)^2 < \infty \}.
\end{aligned}
$$

Proposition 4.8 If $Y(x) \in W$, then

(1) the series

$$\sin x = \sum_{n=1}^{\infty} B^{-1}(n) Y(nx) \tag{4.36}$$

converges unconditionally in the sense of the norm of $L^2[-\pi, \pi]$;

(2) the frequency system $\{Y(nx)\}_{n=1}^{\infty}$ is complete in $L^2_{\text{odd}}[-\pi, +\pi]$.

Proof.

(1) In fact we have

$$\begin{aligned} & \sum_{n=1}^{\infty} B^{-1}(n) Y(nx) \\ = & \sum_{n=1}^{\infty} B^{-1}(n) \sum_{m=1}^{\infty} B(m) \sin(mnx) && (4.37) \\ = & \sum_{l=1}^{\infty} \sum_{mn=l} B^{-1}(n) B(m) \sin lx && (4.38) \\ = & \sin x . \end{aligned}$$

From (4.37) to (4.38), the summation order was changed. But this doesn't matter for the series encountered here are unconditionally convergent in the sense of the norm of $L^2[-\pi, \pi]$. The reason for unconditional convergence is that the series

$$\sum_{l=1}^{\infty} \sum_{mn=l} \mid B^{-1}(n) B(m) \mid \sin lx$$

is convergent, and its norm square

$$\pi \sum_{l=1}^{\infty} \left(\sum_{mn=l} \mid B^{-1}(n) B(m) \mid \right)^2 < \infty. \tag{4.39}$$

(2) From (4.36), it follows that

$$\sin mx = \sum_{n=1}^{\infty} B^{-1}(n) Y(mnx) \tag{4.40}$$

in $L^2_{\text{odd}}[-\pi, +\pi]$ for $m = 1, 2, 3, \cdots$.

Since the sine function system $\{\sin nx\}_{n=1}^{\infty}$ is complete in $L^2_{\text{odd}}[-\pi, +\pi]$, so is the frequency system $\{Y(nx)\}_{n=1}^{\infty}$. □

Definition. The function set Y_Q is defined by

$$Y_Q = \{Y(x): \quad Y(x) = \sum_{n=1}^{\infty} B(n) \sin x, \text{where } B(n) \text{ is quadratically summable and completely multiplicative}\}.$$

If $Y(x) \in Y_Q$, one says that $Y(x)$ is an odd function with quadratically summable and completely multiplicative Fourier coefficients, or an odd WQSCMFC function. Here the term that $B(n)$ is quadratically summable means that

$$\sum_{n=1}^{\infty} B^2(n) < \infty.$$

Proposition 4.9 Y_Q is a subset of W, i.e., $Y_Q \subset W$.

Proof. Since $B(n)$ is completely multiplicative, we have $B^{-1}(n) = \mu(n)B(n)$. It follows that

$$\begin{aligned}
&\sum_{l=1}^{\infty}\left(\sum_{mn=l} | B^{-1}(n)B(m) |\right)^2 \\
&= \sum_{l=1}^{\infty} B^2(l)\left(\sum_{mn=l} | \mu(n) |\right)^2 \\
&\leq \sum_{l=1}^{\infty} B^2(l)d^2(l) \\
&\leq \sum_{l=1}^{\infty} B^2(l) d * d\,(l) \\
&= \left(\sum_{l=1}^{\infty} B^2(l)\right)^4 \\
&< \infty.
\end{aligned}$$

Here $d(n)$ is the divisor function. We have used a relation in Proposition 1.2:

$$d^2\,(n) \;\leq\; d * d\,(n) \;=\; \epsilon^{4*}(n). \tag{4.41}$$

Therefore we obtain that $Y_Q \subset W$. □

Proposition 4.10 If $Y(x) \in Y_Q$, and

$$\psi(x) = \sum_{n=1}^{\infty} b(n) \sin nx$$

is a function in $L^2_{\rm odd}[-\pi, +\pi]$ with

$$\sum_{n=1}^{\infty} b^2(n)d(n) < \infty, \tag{4.42}$$

then $\psi(x)$ can be expressed as a frequency system based on $Y(x)$:

$$\psi(x) = \sum_{n=1}^{\infty} D(n)Y(nx),$$

where

$$D(n) = \sum_{d|n} \mu(\frac{n}{d})B(\frac{n}{d})b(d) = B^{-1} * b\ (n). \tag{4.43}$$

Proof.
This proof is similar to that in Chapter 2. With the help of Chapter 2, readers should have no difficulties in constructing a complete proof except for the use of a new inequality

$$\sum_{j=1}^{\infty} B^2(j)d^3(j) \le \left(\sum_{j=1}^{\infty} B^2(j)\right)^8 < \infty. \tag{4.44}$$

To verify this inequality, one need another relation in Proposition 1.2:

$$d^3\ (n) \ \le\ d * d * d * d\ (n) \ =\ \epsilon^{8*}(n). \tag{4.45}$$

Formula 4.1 If $Y(x) \in Y_{\rm Q}$, we have

$$\langle Y(mx), Y(nx)\rangle = \| Y(x) \|^2 \ B(\frac{mn}{(m,n)^2}), \tag{4.46}$$

for any two positive integers m, n.

Proof. This is easy to verify.

Proposition 4.11 If $Y(x) \in Y_{\rm Q}$, the orthonormalized functions of the frequency system $\{Y(nx)\}$ can be given explicitly by

$$e_n(x) = \frac{1}{\| Y(x) \|} \frac{1}{\sqrt{\beta(n)}} \sum_{d|n} \mu(\frac{n}{d})B(\frac{n}{d})Y(dx), \tag{4.47}$$

where the arithmetical function $\beta(n)$ is defined by

$$\beta(n) = \sum_{d|n} \mu(d) B^2(d). \tag{4.48}$$

Proof.
Since this proof has little difference from that in Chapter 2, we omit it. Obviously the orthonormal system $\{e_n\}_{n=1}^{\infty}$ is an orthonormal basis of $L^2_{\text{odd}}[-\pi, +\pi]$. □

Proposition 4.12 If $Y_1(x) \in Y_Q$, $Y_2(x) \in Y_Q$, and that their Fourier coefficients are $B_1(n)$ and $B_2(n)$, respectively, then the two functions are related by

$$Y_1(x) = \sum_{n=1}^{\infty} D_{12}(n) Y_2(nx), \tag{4.49}$$

$$Y_2(x) = \sum_{n=1}^{\infty} D_{21}(n) Y_1(nx), \tag{4.50}$$

where

$$D_{12}(n) = \sum_{d|n} \mu(\frac{n}{d}) B_2(\frac{n}{d}) B_1(d) = B_1 * B_2^{-1}\,(n), \tag{4.51}$$

$$D_{21}(n) = \sum_{d|n} \mu(\frac{n}{d}) B_1(\frac{n}{d}) B_2(d) = B_2 * B_1^{-1}\,(n). \tag{4.52}$$

Proof. From Proposition 4.10, we can obtain this proposition easily. □

As examples, let us consider several basic waveforms, which are frequently-used and easily-generated in electronics.

Example 1. The I-sawtooth wave

The I-sawtooth wave $Y_{\text{Isa}}(x)$ is a function defined on $(-\infty, +\infty)$ with the period 2π, whose value in one period is given by

$$Y_{\text{Isa}}(x) = \begin{cases} (\pi - x)/2 & 0 < x < 2\pi, \\ 0 & x = 0. \end{cases} \tag{4.53}$$

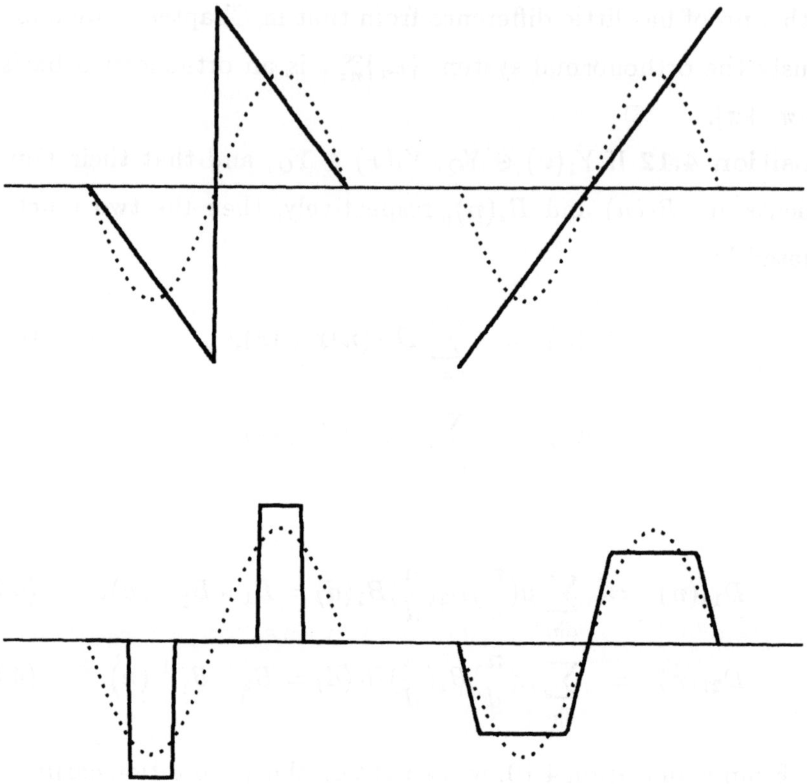

Figure 4.1: The I-sawtooth wave, the II-sawtooth wave, the three-valued function, and the trapezoidal wave.

$$\begin{pmatrix} Y_{\mathrm{Isa}}(x) & Y_{\mathrm{IIsa}}(x) \\ Y(\alpha, a; x) & Y_{\mathrm{tra}}(\alpha, a; x) \end{pmatrix}.$$

See Figure 4.1. Its Fourier series is [3]

$$\begin{aligned} Y_{\text{Isa}}(x) &= \sin x + \frac{1}{2}\sin(2x) + \frac{1}{3}\sin(3x) + \cdots + \frac{1}{n}\sin(nx) + \cdots \\ &= \sum_{n=1}^{\infty} B_{\text{Isa}}(n)\sin(nx). \end{aligned}$$

Its Fourier coefficient

$$B_{\text{Isa}}(n) = 1/n \tag{4.54}$$

is completely multiplicative, so the I-sawtooth wave $Y_{\text{Isa}}(x)$ is an odd WQSCMFC function, i.e., $Y_{\text{Isa}}(x) \in Y_{\text{Q}}$.
Therefore the sine function can be considered as a superposition of I-sawtooth waves with different frequencies:

$$\sin x = Y_{\text{Isa}}(x) - \frac{1}{2}Y_{\text{Isa}}(2x) - \frac{1}{3}Y_{\text{Isa}}(3x) + \cdots + \frac{\mu(n)}{n}Y_{\text{Isa}}(nx) + \cdots. \tag{4.55}$$

The frequency system based on the I-sawtooth

$$Y_{\text{Isa}}(x),\ Y_{\text{Isa}}(2x),\ Y_{\text{Isa}}(3x),\ \cdots,\ Y_{\text{Isa}}(nx),\ \cdots \tag{4.56}$$

is complete in $L^2_{\text{odd}}[-\pi, +\pi]$.
(However whether it is a basis of $L^2_{\text{odd}}[-\pi, +\pi]$ is still unknown.)
Its biorthogonal system is

$$\frac{1}{\pi}\sin x,\ \frac{1}{\pi}(\sin 2x - \frac{1}{2}\sin x),\ \frac{1}{\pi}(\sin 3x - \frac{1}{3}\sin x),\ \cdots,\ \frac{1}{\pi}\sum_{d|n}\frac{d}{n}\mu(\frac{n}{d})\sin dx,\ \cdots,$$

and its orthonormalized system is

$$\sqrt{\tfrac{6}{\pi^3}}Y_{\text{Isa}}(x),\ \sqrt{\tfrac{8}{\pi^3}}(Y_{\text{Isa}}(2x) - \tfrac{1}{2}Y_{\text{Isa}}(x)),\ \sqrt{\tfrac{27}{4\pi^3}}(Y_{\text{Isa}}(3x) - \tfrac{1}{3}Y_{\text{Isa}}(x)),$$

$$\cdots,\sqrt{\tfrac{6}{\pi^3}}\frac{1}{\sqrt{\beta(n)}}\sum_{d|n}\tfrac{d}{n}\mu(\tfrac{n}{d})Y_{\text{Isa}}(dx),\ \cdots,$$

where

$$\beta(n) = \sum_{d|n}\frac{\mu(d)}{d^2}. \tag{4.57}$$

Example 2. The II-sawtooth wave

The II-sawtooth wave $Y_{\mathrm{IIsa}}(x)$ is a function defined on $(-\infty, +\infty)$ with the period 2π (Fig.4.1), whose value in one period is given by

$$Y_{\mathrm{IIsa}}(x) = \begin{cases} x/2 & -\pi < x < \pi, \\ 0 & x = -\pi, \pi. \end{cases} \tag{4.58}$$

Its Fourier series is [3]

$$\begin{aligned} Y_{\mathrm{IIsa}}(x) &= \sin x - \frac{1}{2}\sin(2x) + \frac{1}{3}\sin(3x) + \cdots + \frac{(-1)^{n+1}}{n}\sin(nx) + \cdots \\ &= \sum_{n=1}^{\infty} B_{\mathrm{IIsa}}(n)\sin(nx). \end{aligned}$$

Its Fourier coefficient

$$B_{\mathrm{IIsa}}(n) = \frac{(-1)^{n+1}}{n} = \frac{\theta(n)}{n} \tag{4.59}$$

is not completely multiplicative, so the II-sawtooth wave $Y_{\mathrm{IIsa}}(x)$ is not a WQSCMFC function, i.e., $Y_{\mathrm{IIsa}}(x)$ is not in the set Y_Q. Here θ is the arithmetical function defined by

$$\theta(n) = (-1)^{n+1}. \tag{4.60}$$

Furthermore, it is not difficult to prove that $Y_{\mathrm{IIsa}}(x)$ is not in the set W either (hint: see (4.69)).

Still, the frequency system based on the II-sawtooth

$$Y_{\mathrm{IIsa}}(x),\ Y_{\mathrm{IIsa}}(2x),\ Y_{\mathrm{IIsa}}(3x),\ \cdots,\ Y_{\mathrm{IIsa}}(nx),\ \cdots \tag{4.61}$$

is a complete system in $L^2_{\mathrm{odd}}[-\pi, +\pi]$ (see Proposition 4.13).

Its biorthogonal system is

$$\frac{1}{\pi}\sin x,\ \frac{1}{\pi}(\sin 2x + \frac{1}{2}\sin x),\ \frac{1}{\pi}(\sin 3x - \frac{1}{3}\sin x), \cdots, \frac{1}{\pi}\sum_{d|n}\frac{d}{n}\theta^{-1}(\frac{n}{d})\sin dx, \cdots$$

and its orthonormalized system is

$$\begin{aligned} &\sqrt{\tfrac{6}{\pi^3}}Y_{\mathrm{IIsa}}(x), \quad 4\sqrt{\tfrac{2}{5\pi^3}}(Y_{\mathrm{IIsa}}(2x) + \tfrac{1}{4}Y_{\mathrm{IIsa}}(x)), \\ &\sqrt{\tfrac{27}{4\pi^3}}(Y_{\mathrm{IIsa}}(3x) - \tfrac{1}{3}Y_{\mathrm{IIsa}}(x)), \\ &\sqrt{\tfrac{20}{3\pi^3}}(Y_{\mathrm{IIsa}}(4x) + \tfrac{3}{10}Y_{\mathrm{IIsa}}(2x) + \tfrac{1}{5}Y(x)), \ \cdots. \end{aligned}$$

Here θ^{-1} is the Dirichlet inversion of θ, which is given explicitly by

$$\theta^{-1}(n) = \begin{cases} \mu(n) & l = 0, \\ 2^{l-1}\mu(2k+1) & l = 1, 2, 3, \cdots. \end{cases} \tag{4.62}$$

Here l and k are determined by

$$n = 2^l(2k+1). \tag{4.63}$$

Proposition 4.13 The frequency system based on the II-sawtooth wave, i.e., $\{Y_{\text{IIsa}}(nx)\}_{n=1}^{\infty}$, is a complete system in $L^2_{\text{odd}}[-\pi, +\pi]$, but it is not a basis of $L^2_{\text{odd}}[-\pi, +\pi]$.

Proof.

1. It is complete in $L^2_{\text{odd}}[-\pi, +\pi]$.

Suppose that there exists a function $\psi(x) \in L^2_{\text{odd}}[-\pi, +\pi]$ such that

$$\langle Y_{\text{IIsa}}(nx), \psi(x)\rangle = 0 \quad \text{for } n = 1, 2, 3, \cdots. \tag{4.64}$$

From the definitions of the two sawtooth waves, we have the relation

$$Y_{\text{IIsa}}(x) = Y_{\text{Isa}}(x) - Y_{\text{Isa}}(2x). \tag{4.65}$$

Thus we obtain that

$$\langle Y_{\text{Isa}}(nx), \psi(x)\rangle = \langle Y_{\text{Isa}}(2nx), \psi(x)\rangle \quad \text{for } n = 1, 2, 3, \cdots. \tag{4.66}$$

Furthermore, for an arbitrary positive integer l we have

$$\langle Y_{\text{Isa}}(lx), \psi(x)\rangle = \langle Y_{\text{Isa}}(2lx), \psi(x)\rangle = \cdots = \langle Y_{\text{Isa}}(2^m lx), \psi(x)\rangle = \cdots$$

where $m = 0, 1, 2, \cdots$.

Due to Proposition 4.2, $Y_{\text{Isa}}(nx)$ converges weakly to zero, and we have

$$\langle Y_{\text{Isa}}(lx), \psi(x)\rangle = 0. \tag{4.67}$$

Since $\{Y_{\text{Isa}}(nx)\}_{n=1}^{\infty}$ is complete, we have

$$\psi(x) = 0. \tag{4.68}$$

This completes the proof about the completeness.

2. It is not a basis of $L^2_{\text{odd}}[-\pi, +\pi]$.

If it is a basis of $L^2_{\text{odd}}[-\pi, +\pi]$, by Proposition 4.7, $\sin x$ can be expressed as a frequency series:

$$\sin x = Y_{\text{IIsa}}(x) + \frac{1}{2}Y_{\text{IIsa}}(2x) - \frac{1}{3}Y_{\text{IIsa}}(3x) + \cdots + \frac{\theta^{-1}(n)}{n}Y_{\text{IIsa}}(nx) + \cdots.$$

But this is impossible, because $\frac{\theta^{-1}(n)}{n}$ does not converge to zero. In fact, for $n = 2^m, m = 1, 2, 3, \cdots$, we always have

$$\frac{\theta^{-1}(n)}{n} = \frac{1}{2}. \tag{4.69}$$

Therefore it is not a basis of $L^2_{\text{odd}}[-\pi, +\pi]$.

Example 3. a periodic three-valued function

Suppose that $Y(\alpha, a; x)$ is a three-valued function of the period 2π (Fig.4.1), the value in one period of which is given by

$$Y(\alpha, a; x) = \begin{cases} 0 & -\pi < x < -\pi + \alpha,\ -\alpha < x < \alpha,\ \pi - \alpha < x < \pi, \\ -a & -\pi + \alpha < x < -\alpha, \\ a & \alpha < x < \pi - \alpha, \end{cases}$$

where $a > 0$ and $0 \le \alpha < \pi/2$.

The Fourier series of $Y(\alpha, a; x)$ is [3]

$$\begin{aligned} Y(\alpha, a; x) &= \frac{4a}{\pi}(\cos\alpha \sin x + \frac{1}{3}\cos 3\alpha \sin 3x + \frac{1}{5}\cos 5\alpha \sin 5x + \cdots) \\ &= \sum_{n=1}^{\infty} B(n) \sin nx, \end{aligned}$$

where the Fourier coefficients

$$B(n) = \begin{cases} \frac{4a}{\pi}\frac{\cos n\alpha}{n} & n = 1, 3, 5, \cdots, \\ 0 & n = 2, 4, 6, \cdots. \end{cases} \tag{4.70}$$

Proposition 4.14 $Y(\alpha, a; x)$ is an odd WQSCMFC function, i.e. $Y(\alpha, a; x) \in Y_{\text{Q}}$ if and only if

$$(0) \quad \alpha = 0, \qquad a = \frac{\pi}{4}; \tag{4.71}$$

$$(1) \quad \alpha = \frac{\pi}{4}, \qquad a = \frac{\sqrt{2}\pi}{4}; \tag{4.72}$$

or

$$(2) \quad \alpha = \frac{\pi}{6}, \qquad a = \frac{\sqrt{3}\pi}{6}. \tag{4.73}$$

Proof.

It is easy to verify that in the case (0), (1) or (2), the function $Y(\alpha, a; x)$ is an odd WQSCMFC function indeed.

Now let us prove that except the three cases $Y(\alpha, a; x)$ is not an odd WQSCMFC function.

If $Y(\alpha, a; x)$ is an odd WQSCMFC function, then its Fourier coefficients

$$B(n) = \begin{cases} \frac{4a}{\pi}\frac{\cos n\alpha}{n} & n = 1, 3, 5, \cdots, \\ 0 & n = 2, 4, 6, \cdots. \end{cases} \tag{4.74}$$

are a completely multiplicative arithmetical function. From $B(1) = 1$ it follows that

$$a = \frac{\pi}{4}\frac{1}{\cos\alpha}.$$

Form $B^2(3) = B(9)$ it follows that

$$\left(\frac{\cos 3\alpha}{\cos\alpha}\right)^2 = \frac{\cos 9\alpha}{\cos\alpha}. \tag{4.75}$$

The equation (4.75) has three solutions in $[0, \frac{\pi}{2})$:

$$\alpha = 0, \ \alpha = \frac{\pi}{4}, \ \text{or} \ \alpha = \frac{\pi}{6}.$$

The associated values of a are

$$a = \frac{\pi}{4}, \ a = \frac{\sqrt{2}\pi}{4}, \ \text{or} \ a = \frac{\sqrt{3}\pi}{6}.$$

Thus there exist at most three WQSCMFC functions among all $Y(\alpha, a; x)$ with $a > 0$ and $0 \le \alpha < \pi/2$.

This completes the proof. □.

For convenience, we call the three functions

$$Y(0, \frac{\pi}{4}; x) = Y_{\text{sq}}(x),$$

$$Y(\frac{\pi}{4}, \frac{\sqrt{2}\pi}{4}; x) = Y_{\pi/4}(x),$$

$$Y(\frac{\pi}{6}, \frac{\sqrt{3}\pi}{6}; x) = Y_{\pi/6}(x),$$

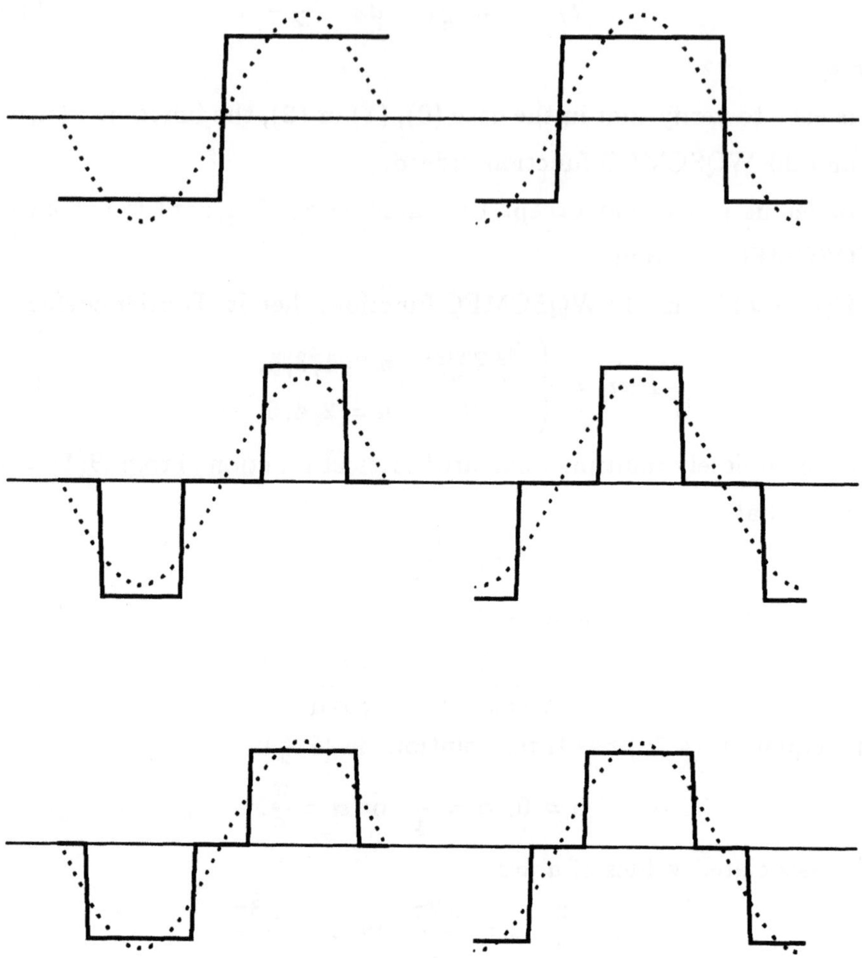

Figure 4.2: The odd and even square waves, $\frac{\pi}{4}$ three-valued functions, and $\frac{\pi}{6}$ three-valued functions.

$$\begin{pmatrix} Y_{sq}(x) & X_{sq}(x) \\ Y_{\pi/4}(x) & X_{\pi/4}(x) \\ Y_{\pi/6}(x) & X_{\pi/6}(x) \end{pmatrix}.$$

odd square wave, $\frac{\pi}{4}$ three-valued function, and $\frac{\pi}{6}$ three-valued function, respectively, which are common in electronics. See Figure 4.2.

Here are three open questions:

1) In the case (0), (1) or (2), is the frequency system $\{Y(\alpha, a; nx)\}$ a basis of $L^2_{\text{odd}}[-\pi, +\pi]$?

2) Except the cases (0), (1) and (2), is the frequency system $\{Y(\alpha, a; nx)\}$ a complete system in $L^2_{\text{odd}}[-\pi, +\pi]$?

3) Beyond the three-valued functions, is there any other odd piecewise constant function $Y(x) \in Y_{\text{Q}}$? (such as five-valued functions.)

By Proposition 4.12, the sawtooth wave $Y_{\text{lsa}}(x)$ and the odd square waves $Y_{\text{sq}}(x)$ are related by

$$Y_{\text{lsa}}(x) = Y_{\text{sq}}(x) + \frac{1}{2}Y_{\text{sq}}(2x) + \frac{1}{4}Y_{\text{sq}}(4x) + \cdots + \frac{1}{2^m}Y_{\text{sq}}(2^m) + \cdots,$$

and

$$Y_{\text{sq}}(x) = Y_{\text{lsa}}(x) - \frac{1}{2}Y_{\text{lsa}}(2x). \tag{4.76}$$

4.4 An Unconditional Basis in $L^2_{\text{odd}}[-\pi, +\pi]$

We shall discuss when a frequency system is an unconditional basis of $L^2_{\text{odd}}[-\pi, +\pi]$.

Definition. A function set B is defined by

$$B = \{\, Y(x) \;:\; Y(x) = \sum_{n=1}^{\infty} B(n)\sin nx, \text{ where } B(n) \text{ satisfies } \sum_{n=1}^{\infty} |\, B(n) \,| < \infty, \text{ and } \sum_{n=1}^{\infty} |\, B^{-1}(n) \,| < \infty \,\}.$$

Obviously, B is a subset of W.

Proposition 4.15 If $Y(x) \in B$, then

(1) the series

$$\sin x = \sum_{n=1}^{\infty} B^{-1}(n) Y(nx) \tag{4.77}$$

converges absolutely and uniformly in $[-\pi, \pi]$;
(2) the frequency system $\{Y(nx)\}$ is an unconditional basis of $L^2_{\text{odd}}[-\pi, +\pi]$.

Proof.

(1) In fact we have

$$\begin{aligned}
&\sum_{n=1}^{\infty} B^{-1}(n)Y(nx)\\
&= \sum_{n=1}^{\infty} B^{-1}(n) \sum_{m=1}^{\infty} B(m)\sin(mnx) && (4.78)\\
&= \sum_{l=1}^{\infty} \sum_{mn=l} B^{-1}(n)B(m)\sin lx && (4.79)\\
&= \sin x.
\end{aligned}$$

Since

$$\begin{aligned}
&\sum_{n=1}^{\infty} | B^{-1}(n) | \sum_{m=1}^{\infty} | B(m)\sin(mnx) |\\
&\leq \sum_{n=1}^{\infty} | B^{-1}(n) | \sum_{m=1}^{\infty} | B(m) |\\
&< \infty,
\end{aligned}$$

by Weierstrass M-test, we obtain that the series (4.77) converges absolutely and uniformly in $[-\pi, \pi]$.

(2) Suppose that $\psi(x) \in L^2_{\text{odd}}[-\pi, +\pi]$ is an arbitrary function and that $b(n)$ denotes its Fourier coefficients. In fact, we have

$$\begin{aligned}
\psi(x) &= \sum_{n=1}^{\infty} b(n)\sin nx && (4.80)\\
&= \sum_{n=1}^{\infty} b(n) \sum_{l=1}^{\infty} B^{-1}(l)Y(nlx) && (4.81)\\
&= \sum_{n=1}^{\infty} \left(\sum_{d|n} b(d)B^{-1}\left(\frac{n}{d}\right)\right)Y(nx) && (4.82)\\
&= \sum_{n=1}^{\infty} D(n)Y(nx), && (4.83)
\end{aligned}$$

where

$$D(n) = \sum_{d|n} b(d)B^{-1}(\frac{n}{d}) = B^{-1} * b\,(n). \tag{4.84}$$

From (4.81) to (4.82), the summation orders were changed. But this doesn't matter because all the following series

$$\begin{aligned}
&\sum_{n=1}^{\infty} b(n) \sum_{l=1}^{\infty} B^{-1}(l) Y(nlx) \\
= &\sum_{n=1}^{\infty} b(n) \sum_{l=1}^{\infty} B^{-1}(l) \sum_{m=1}^{\infty} B(m) \sin(nlmx) \\
= &\sum_{k=1}^{\infty} (\sum_{nlm=k} b(n) B^{-1}(l) B(m)) \sin(kx)
\end{aligned}$$

are unconditionally convergent in $L^2[-\pi,\pi]$. The reason for the unconditional convergence is that the series

$$\sum_{k=1}^{\infty} (\sum_{nlm=k} \mid b(n)B^{-1}(l)B(m) \mid) \sin(kx) \tag{4.85}$$

is convergent, the norm of which is

$$\begin{aligned}
&\| \sum_{k=1}^{\infty} (\sum_{nlm=k} \mid b(n)B^{-1}(l)B(m) \mid) \sin(kx) \ \| \\
= &\| \sum_{m=1}^{\infty} \mid B(m) \mid \sum_{l=1}^{\infty} \mid B^{-1}(l) \mid \sum_{n=1}^{\infty} \mid b(n) \mid \sin(nlmx) \ \| \\
\leq &\sum_{m=1}^{\infty} \mid B(m) \mid \sum_{l=1}^{\infty} \mid B^{-1}(l) \mid \| \sum_{n=1}^{\infty} \mid b(n) \mid \sin(nlmx) \ \| \\
= &\sum_{m=1}^{\infty} \mid B(m) \mid \sum_{l=1}^{\infty} \mid B^{-1}(l) \mid \sqrt{\pi \sum_{n=1}^{\infty} b^2(n)} \\
< &\ \infty.
\end{aligned}$$

We can obtain the uniqueness of the series (4.83) by the consequence of Proposition 4.1.

Therefore $\{Y(nx)\}$ is an unconditional basis of $L^2[-\pi,\pi]$.

This completes the proof. □

Example. The function

$$Y(x) = \sin x - \frac{1}{2^2}\sin 2x + \frac{1}{3^2}\sin 3x + \cdots + \frac{(-1)^{n+1}}{n^2}\sin nx + \cdots,$$

is an element of B, since $B(n)$ satisfy that

$$B(1) = 1, \quad \sum_{n=1}^{\infty} \mid B(n) \mid = \frac{\pi^2}{6}, \quad \text{and} \quad \sum_{n=1}^{\infty} \mid B^{-1}(n) \mid = \frac{18}{\pi^2}.$$

Definition. The function set Y_A is defined by

$$Y_A = \{Y(x) \ : \ Y(x) = \sum_{n=1}^{\infty} B(n)\sin nx, \text{ where } B(n) \text{ is absolutely summable and completely multiplicative}\}.$$

If $Y(x) \in Y_A$, one says that $Y(x)$ is an odd function with absolutely summable and completely multiplicative Fourier coefficients, or an odd WASCMFC function. (In Chapter 3 we write Y simply.)

Proposition 4.16 Y_A is a subset of B, i.e. $Y_A \subset B$.

Proof. Since $B(n)$ is completely multiplicative, we have

$$\sum_{n=1}^{\infty} \mid B^{-1}(n) \mid = \sum_{n=1}^{\infty} \mid \mu(n)B(n) \mid \leq \sum_{n=1}^{\infty} \mid B(n) \mid < \infty. \tag{4.86}$$

Therefore $Y_A \subset B$. Additionally, we have $Y_A \subset Y_Q$.

Example. Trapezoidal wave

The trapezoidal wave $Y_{tra}(a, \alpha; x)$ is a function of the period 2π (Fig.4.1). Its value in one period is given by

$$Y_{tra}(\alpha, a; x) = \begin{cases} ax/\alpha & \text{for } -\alpha \leq x \leq \alpha, \\ a & \text{for } \alpha \leq x \leq \pi - \alpha, \\ -a & \text{for } -\pi + \alpha \leq x \leq -\alpha, \\ a(\pi - x)/\alpha & \text{for } \pi - \alpha \leq x \leq \pi, \\ -a(\pi + x)/\alpha & \text{for } -\pi \leq x \leq -\pi + \alpha, \end{cases} \tag{4.87}$$

where

$$0 < \alpha \leq \pi/2, \ a > 0. \tag{4.88}$$

Its Fourier series is [3]

$$\begin{aligned} Y_{\text{tra}}(\alpha, a; x) &= \frac{4a}{\pi\alpha}(\sin\alpha\sin x + \frac{1}{3^2}\sin 3\alpha\sin 3x + \frac{1}{5^2}\sin 5\alpha\sin 5x + \cdots) \\ &= \sum_{n=1}^{\infty} B(n)\sin nx. \end{aligned}$$

where

$$B(n) = \begin{cases} \frac{4a}{\pi\alpha}\frac{1}{n^2}\sin n\alpha & n = 1, 3, 5, \cdots, \\ 0 & n = 2, 4, 6, \cdots. \end{cases} \tag{4.89}$$

Proposition 4.17 The function $Y_{\text{tra}}(\alpha, a; x)$ is an odd WASCMFC function if and only if

$$(0) \quad \alpha = \frac{\pi}{2}, \qquad a = \frac{\pi^2}{8}, \tag{4.90}$$

$$(1) \quad \alpha = \frac{\pi}{4}, \qquad a = \frac{\sqrt{2}\pi^2}{16}, \tag{4.91}$$

or

$$(2) \quad \alpha = \frac{\pi}{3}, \qquad a = \frac{\sqrt{3}\pi^2}{18}. \tag{4.92}$$

Proof. This proof is similar to that of Proposition 4.14. □

For convenience, we call the three functions

$$\begin{aligned} &Y_{\text{tra}}(\frac{\pi}{2}, \frac{\pi^2}{8}; x) = Y_{\text{tri}}(x) \\ &Y_{\text{tra}}(\frac{\pi}{4}, \frac{\sqrt{2}\pi^2}{16}; x) = Y_{\pi/4\text{tra}}(x) = Y_{\text{Itra}}(x) \\ &Y_{\text{tra}}(\frac{\pi}{3}, \frac{\sqrt{3}\pi^2}{18}; x) = Y_{\pi/3\text{tra}}(x) = Y_{\text{IItra}}(x) \end{aligned}$$

odd triangular wave, $\frac{\pi}{4}$ trapezoidal wave (I-trapezoidal wave), and $\frac{\pi}{3}$ trapezoidal wave (II-trapezoidal wave), respectively, which were discussed carefully in Chapter 3.

Proposition 4.18 If the Fourier coefficients $B(n)$ of $Y(x) \in L^2_{\text{odd}}[-\pi, +\pi]$ satisfy that

$$B(1) = 1 \quad \text{and} \quad \sum_{n=2}^{\infty} | B(n) | < 1, \tag{4.93}$$

then $Y(x) \in B$.

Proof.

In fact we have

$$\begin{aligned}
& \left(\sum_{n=1}^{\infty} B(n)\right)^{-1} \\
= & \left(1+\sum_{n=2}^{\infty} B(n)\right)^{-1} \\
= & 1+(-1)\sum_{n=2}^{\infty} B(n)+\cdots+(-1)^m\left(\sum_{n=2}^{\infty} B(n)\right)^m+\cdots \\
= & 1-B(2)-B(3)+[B(2)B(2)-B(4)]-B(5) \\
& +[2B(2)B(3)-B(6)]-B(7)+\cdots \\
= & \sum_{n=1}^{\infty} B^{-1}(n).
\end{aligned}$$

Therefore we obtain that

$$\begin{aligned}
& \sum_{n=1}^{\infty} \mid B^{-1}(n) \mid \\
\leq & 1+\sum_{n=2}^{\infty} \mid B(n) \mid +\cdots+\left(\sum_{n=2}^{\infty} \mid B(n) \mid\right)^m+\cdots \\
= & \frac{1}{1-\sum_{n=2}^{\infty}|B(n)|} \\
< & \infty.
\end{aligned}$$

This completes the proof. □

Proposition 4.19 For $\pi/2 \geq \alpha > \arcsin(\frac{\pi^2}{8}-1)$, the frequency system $\{Y_{\mathrm{tra}}(\alpha, a; nx)\}$ is an unconditional basis of $L^2_{\mathrm{odd}}[-\pi,+\pi]$.

Proof. It does not matter taking

$$a = \frac{\pi}{4}\frac{\alpha}{\sin\alpha}. \tag{4.94}$$

The Fourier coefficients of $Y_{\mathrm{tra}}(\alpha, \frac{\pi}{4}\frac{\alpha}{\sin\alpha}; x)$ is

$$\begin{aligned}
Y_{\mathrm{tra}}\left(\alpha, \frac{\pi}{4}\frac{\alpha}{\sin\alpha}; x\right) &= \sin x + \frac{1}{3^2}\frac{\sin 3\alpha}{\sin\alpha}\sin 3x + \frac{1}{5^2}\frac{\sin 5\alpha}{\sin\alpha}\sin 5x + \cdots \\
&= \sum_{n=1}^{\infty} B(n)\sin nx.
\end{aligned}$$

Obviously we have $B(1) = 1$. Also we have

$$\begin{aligned}
& \sum_{n=2}^{\infty} \mid B(n) \mid \\
= & \frac{1}{\sin\alpha}\left(\frac{|\sin 3\alpha|}{3^2} + (\frac{|\sin 5\alpha|}{5^2} + \cdots\right) \\
\leq & \frac{1}{\sin\alpha}\left(\frac{1}{3^2} + \frac{1}{5^2} + \cdots\right) \\
= & \frac{1}{\sin\alpha}(\frac{\pi^2}{8} - 1) \\
< & 1.
\end{aligned}$$

By Proposition 4.18, we obtain this proposition. □

Here we have

$$\arcsin(\frac{\pi^2}{8} - 1) \doteq \arcsin(0.23370) \doteq 13^\circ 31'. \tag{4.95}$$

An open question is: what about $0 \leq \alpha \leq \arcsin(\frac{\pi^2}{8} - 1)$?

4.5 A Combinative Frequency System in $L^2[-\pi, \pi]$

Since the case of the even function subspace $L^2_{\text{even}}[-\pi, \pi]$ is similar to the case of the odd function subspace $L^2_{\text{odd}}[-\pi, +\pi]$, let us return directly to the whole space $L^2[-\pi, \pi]$ and draw our general conclusions.

Suppose that

$$X(x) = \sum_{n=1}^{\infty} A(n) \cos nx \tag{4.96}$$

is a given function in $L^2_{\text{even}}[-\pi, +\pi]$, and

$$Y(x) = \sum_{n=1}^{\infty} B(n) \sin nx \tag{4.97}$$

is a given function in $L^2_{\text{odd}}[-\pi, +\pi]$.

Generally, if the frequency system

$$Y(x), Y(2x), Y(3x), \cdots, Y(nx), \cdots \tag{4.98}$$

is a complete system (an unconditional basis) in $L^2_{\text{odd}}[-\pi, +\pi]$, and the frequency system

$$X(x), X(2x), X(3x), \cdots, X(nx), \cdots \tag{4.99}$$

is a complete system (an unconditional basis) in $L^2_{\text{even}}[-\pi, +\pi]$ as well, then the combinative frequency system

$$1, X(x), Y(x), X(2x), Y(2x), \cdots, X(nx), Y(nx), \cdots \tag{4.100}$$

is a complete system (an unconditional basis) in $L^2[-\pi, \pi]$.

If the combinative frequency system (4.100) is a complete system in $L^2[-\pi, \pi]$, then its orthonormalized system is an orthonormal basis of $L^2[-\pi, \pi]$, and any function $f(x) \in L^2[-\pi, \pi]$ can be approximated by a linear combination of finite functions in (4.100) with an arbitrary mean-square error.

If the combinative frequency system (4.100) is an unconditional basis, then any function $f(x) \in L^2[-\pi, \pi]$ can be expressed as a combinative frequency series based on $X(x)$ and $Y(x)$:

$$f(x) = C_0 + \sum_{n=1}^{\infty} C(n)X(nx) + D(n)Y(nx), \tag{4.101}$$

where the coefficients $C_0, C(n)$ and $D(n)$ can be determined by

$$\begin{aligned} C_0 &= \frac{1}{2\pi} \int_{-\pi}^{\pi} f(x)dx, \\ C(n) &= \int_{-\pi}^{\pi} f(x)h_n(x)dx, \\ D(n) &= \int_{-\pi}^{\pi} f(x)g_n(x)dx. \end{aligned} \tag{4.102}$$

Here

$$\begin{aligned} h_n(x) &= \frac{1}{\pi} \sum_{d|n} A^{-1}(\frac{n}{d}) \cos dx \\ g_n(x) &= \frac{1}{\pi} \sum_{d|n} B^{-1}(\frac{n}{d}) \sin dx \end{aligned}$$

are the biorthogonal functions of $X(nx)$ and $Y(nx)$, respectively.

Speaking concretely, the complete combinative frequency systems of practical importance are

1. the square wave system

$$1, X_{sq}(x), Y_{sq}(x), X_{sq}(2x), Y_{sq}(2x), \cdots, X_{sq}(nx), Y_{sq}(nx), \cdots,$$

2. the $\frac{\pi}{4}$ three-valued function system

$$1, X_{\pi/4}(x), Y_{\pi/4}(x), X_{\pi/4}(2x), Y_{\pi/4}(2x), \cdots, X_{\pi/4}(nx), Y_{\pi/4}(nx), \cdots,$$

3. the $\frac{\pi}{6}$ three-valued function system

$$1, X_{\pi/6}(x), Y_{\pi/6}(x), X_{\pi/6}(2x), Y_{\pi/6}(2x), \cdots, X_{\pi/6}(nx), Y_{\pi/6}(nx), \cdots,$$

and so on. Here the functions

$$X_{sq}(x) = Y_{sq}(x + \frac{\pi}{2})$$
$$X_{\pi/4}(x) = Y_{\pi/4}(x + \frac{\pi}{2})$$
$$X_{\pi/6}(x) = Y_{\pi/6}(x + \frac{\pi}{2})$$

are called even square wave, $\frac{\pi}{4}$ three-valued function, $\frac{\pi}{6}$ three-valued function, respectively. See Figure 4.2.

The unconditional bases of the practical importance are

1. the triangular wave basis

$$1, X_{tri}(x), Y_{tri}(x), X_{tri}(2x), Y_{tri}(2x), \cdots, X_{tri}(nx), Y_{tri}(nx), \cdots,$$

2. the I-trapezoidal wave basis

$$1, X_{Itra}(x), Y_{Itra}(x), X_{Itra}(2x), Y_{Itra}(2x), \cdots, X_{Itra}(nx), Y_{Itra}(nx), \cdots,$$

3. the II-trapezoidal wave basis

$$1, X_{IItra}(x), Y_{IItra}(x), X_{IItra}(2x), Y_{IItra}(2x), \cdots, X_{IItra}(nx), Y_{IItra}(nx), \cdots,$$

and so on.

In fact, for $\alpha > 13^o 31'$, a general trapezoidal wave system

$$1, \{X_{\text{tra}}(\alpha, a; nx), Y_{\text{tra}}(\alpha, a; nx),\}$$

is an unconditional basis of $L^2[-\pi, \pi]$, where the function

$$X_{\text{tra}}(\alpha, a; x) = Y_{\text{tra}}(\alpha, a; x + \frac{\pi}{2}) \tag{4.103}$$

is called even trapezoidal wave.

4.6 A Frequency Transform in $L^2(R)$

In this section, we shall introduce in $L^2(R)$ the frequency transform based on general periodic functions, which is a generalization of Fourier transform based on sine-cosine functions.

$L^2(-\infty, +\infty)$ or $L^2(R)$ denotes the space of quadratically integrable functions on $R = (-\infty, +\infty)$. It is well known that $L^2(R)$ can be decomposed into an even function subspace $L^2_{\text{even}}(R)$ and an odd function subspace $L^2_{\text{odd}}(R)$:

$$L^2(R) = L^2_{\text{even}}(R) \bigoplus L^2_{\text{odd}}(R). \tag{4.104}$$

By Plancherel's theory [7], Fourier transform $\mathbf{F}$ and its inverse transform $\mathbf{F}^{-1}$ are two bounded linear operators from $L^2(R)$ to itself, i.e., $\mathbf{F}$, $\mathbf{F}^{-1} \in B(L^2(R))$. $B(L^2(R))$ denotes the Banach space of all the bounded linear operators from $L^2(R)$ to itself.

That is to say, for any two functions $f(x), \hat{f}(\omega) \in L^2(R)$, we have

$$\hat{f}(\omega) = (\mathbf{F}f)(\omega) = \frac{1}{2\pi}\int_{-\infty}^{+\infty} f(x)e^{-i\omega x}dx, \tag{4.105}$$

$$\Longleftrightarrow$$

$$f(x) = (\mathbf{F}^{-1}\hat{f})(x) = \int_{-\infty}^{+\infty} \hat{f}(\omega)e^{i\omega x}d\omega. \tag{4.106}$$

Strictly speaking, when the integrals in (4.105) and (4.106) do not exist, $\int_{-\infty}^{+\infty}$ should be considered to be $\lim\limits_{N\to\infty}\int_{-N}^{N}$, where the limit is in the sense of the norm in $L^2(R)$, see [7].

In the even function subspace $L^2_{\text{even}}(R)$, Fourier transform $\mathbf{F}$ and its inversion $\mathbf{F}^{-1}$ become Fourier cosine transform $\mathbf{F}_{\cos}$ and its inversion $\mathbf{F}^{-1}_{\cos}$. That is to say, for any two functions $f_{\text{even}}(x), a(\omega) \in L^2_{\text{even}}(R)$, we have

$$a(\omega) = (\mathbf{F}_{\cos} f_{\text{even}})(\omega) = \frac{1}{2\pi}\int_{-\infty}^{+\infty} f_{\text{even}}(x)\cos(\omega x)dx,$$

$$\Longleftrightarrow$$

$$f_{\text{even}}(x) = (\mathbf{F}^{-1}_{\cos} a)(x) = \int_{-\infty}^{+\infty} a(\omega)\cos(\omega x)d\omega.$$

In the odd function subspace $L^2_{\text{odd}}(R)$, there exist similar Fourier sine transform and its inversion.

Next let us introduce the frequency transform based on general periodic functions step by step.

In this section we suppose that $X(x)$, $Y(x) \in L^2[-\pi,\pi]$ are given even and odd function with period 2π. Their Fourier series are

$$\begin{aligned} X(x) &= \sum_{n=1}^{\infty} A(n)\cos nx, \\ Y(x) &= \sum_{n=1}^{\infty} B(n)\sin nx, \end{aligned}$$

and their Fourier coefficients satisfy

$$\sum_{n=1}^{\infty} | A(n) | \frac{1}{\sqrt{n}} < \infty, \qquad \sum_{n=1}^{\infty} | A^{-1}(n) | \frac{1}{\sqrt{n}} < \infty,$$

$$\sum_{n=1}^{\infty} | B(n) | \frac{1}{\sqrt{n}} < \infty, \qquad \sum_{n=1}^{\infty} | B^{-1}(n) | \frac{1}{\sqrt{n}} < \infty.$$

Definition. The functions $\widetilde{X}(x)$ and $\widetilde{Y}(x)$ are called the dual functions of $X(x)$ and $Y(x)$, respectively, which are defined on $(-\infty,\infty)$ by

$$\widetilde{X}(x) = \frac{1}{2\pi}\sum_{n=1}^{\infty} A^{-1}(n)\frac{1}{n}\cos(\frac{x}{n}), \tag{4.107}$$

$$\widetilde{Y}(x) = \frac{1}{2\pi}\sum_{n=1}^{\infty} B^{-1}(n)\frac{1}{n}\sin(\frac{x}{n}). \quad (4.108)$$

Proposition 4.20 The functions $\widetilde{X}(x)$ and $\widetilde{Y}(x)$ are bounded and infinitely differentiable in $(-\infty, +\infty)$.

Proof. We have

$$\begin{aligned} |\widetilde{X}(x)| &\le \frac{1}{2\pi}\sum_{n=1}^{\infty} |A^{-1}(n)\frac{1}{n}\cos(\frac{x}{n})| \\ &\le \frac{1}{2\pi}\sum_{n=1}^{\infty} |A^{-1}(n)|\frac{1}{n}. \end{aligned}$$

Therefore the function $\widetilde{X}(x)$ is bounded. By Weierstrass M-test, the series (4.107) converges absolutely and uniformly in $(-\infty, \infty)$. By properties of a uniformly convergent series, the function $\widetilde{X}(x)$ is infinitely differentiable.

The case of the function $\widetilde{Y}(x)$ is similar.

Definition. The operator $\mathbf{T}(n) : L^2(R) \to L^2(R)$ is defined by

$$\mathbf{T}(n)f(x) = f(\frac{x}{n}), \quad (4.109)$$

where $f(x) \in L^2(R)$ and n is any positive integer.

Proposition 4.21

1.The operator $\mathbf{T}(n)$ is a bounded linear operator, i.e. $\mathbf{T}(n) \in B(L^2(R))$, and its norm $\| \mathbf{T}(n) \| = \sqrt{n}$.

2.For any two positive integers m and n, we have

$$\mathbf{T}(m)\mathbf{T}(n) = \mathbf{T}(mn), \quad (4.110)$$

that is to say $\mathbf{T}$(n) is completely multiplicative.

Proof. This proposition is clear from the definition. □

Proposition 4.22 The operator series

$$\mathbf{S} = \sum_{n=1}^{\infty} A(n)\frac{1}{n}\mathbf{T}(n) \quad (4.111)$$

$$\mathbf{S}^{-1} = \sum_{n=1}^{\infty} A^{-1}(n)\frac{1}{n}\mathbf{T}(n) \quad (4.112)$$

converge absolutely in the sense of the norm in $B(L^2(R))$, and are mutually inverse.

Proof.

Because

$$\sum_{n=1}^{\infty} \| A(n)\frac{1}{n}\mathbf{T}(n) \| \leq \sum_{n=1}^{\infty} | A(n) | \frac{1}{n} \| \mathbf{T}(n) \| = \sum_{n=1}^{\infty} | A(n) | \frac{1}{\sqrt{n}} < \infty,$$

$$\sum_{n=1}^{\infty} \| A^{-1}(n)\frac{1}{n}\mathbf{T}(n) \| \leq \sum_{n=1}^{\infty} | A^{-1}(n) | \frac{1}{n} \| \mathbf{T}(n) \| = \sum_{n=1}^{\infty} | A^{-1}(n) | \frac{1}{\sqrt{n}} < \infty,$$

the two operator series converge absolutely in the sense of the norm in $B(L^2(R)$. Thus $\mathbf{S}$ and $\mathbf{S}^{-1}$ are two bounded linear operators defined on $L^2(R)$, i.e. $\mathbf{S},\ \mathbf{S}^{-1} \in B(L^2(R))$.

Furthermore, we have

$$\begin{aligned}
&\mathbf{S}\mathbf{S}^{-1} = \mathbf{S}^{-1}\mathbf{S} \\
&= \sum_{n=1}^{\infty} A(n)\frac{1}{n}\mathbf{T}(n) \sum_{m=1}^{\infty} A^{-1}(m)\frac{1}{m}\mathbf{T}(m) \\
&= \sum_{k=1}^{\infty} \frac{1}{k}\mathbf{T}(k) \sum_{n|k} A(n)A^{-1}(\frac{k}{n}) \\
&= \mathbf{T}(1) = \mathbf{I},
\end{aligned}$$

where $\mathbf{I}$ is the identical operator. That is to say, the operators $\mathbf{S}$ and $\mathbf{S}^{-1}$ are mutually inverse. □

Proposition 4.23 For any two functions $f_{\text{even}}(x), C(\omega) \in L^2_{\text{even}}(R)$, we have

$$C(\omega) = (\mathbf{W}_{\text{even}} f_{\text{even}})(\omega) = \int_{-\infty}^{+\infty} f_{\text{even}}(x)\widetilde{X}(\omega x)dx,$$

$$\Longleftrightarrow$$

$$f_{\text{even}}(x) = (\mathbf{W}^{-1}_{\text{even}} C)(x) = \int_{-\infty}^{+\infty} C(\omega)X(\omega x)d\omega.$$

Here $\mathbf{W}_{\text{even}} : L^2_{\text{even}}(R) \to L^2_{\text{even}}(R)$ is called the frequency transform based on the even period function $X(x)$, and $\mathbf{W}^{-1}_{\text{even}} : L^2_{\text{even}}(R) \to L^2_{\text{even}}(R)$ is its inverse.

Proof. Let us consider the case $\Longrightarrow$.

First we have

$$\begin{aligned}
C(\omega) &= (\mathbf{W}_{\text{even}} f_{\text{even}})(\omega) \\
&= \int_{-\infty}^{+\infty} f_{\text{even}}(x) \widetilde{X}(\omega x) dx \\
&= \frac{1}{2\pi} \int_{-\infty}^{+\infty} f_{\text{even}}(x) \sum_{n=1}^{\infty} A^{-1}(n) \frac{1}{n} \cos(\frac{x\omega}{n}) dx, \\
&= \frac{1}{2\pi} \sum_{n=1}^{\infty} A^{-1}(n) \frac{1}{n} \int_{-\infty}^{+\infty} f_{\text{even}}(x) \cos(\frac{x\omega}{n}) dx \\
&= \frac{1}{2\pi} \sum_{n=1}^{\infty} A^{-1}(n) \frac{1}{n} \mathbf{T}_n \int_{-\infty}^{+\infty} f_{\text{even}}(x) \cos(x\omega) dx \\
&= \sum_{n=1}^{\infty} A^{-1}(n) \frac{1}{n} \mathbf{T}_n \mathbf{F}_{\cos} f_{\text{even}}(x) \\
&= \mathbf{S}^{-1} \mathbf{F}_{\cos} f_{\text{even}}(x).
\end{aligned}$$

In one word, in $L^2_{\text{even}}(R)$ the operator $\mathbf{W}_{\text{even}}$ can be decomposed into two operators:

$$\mathbf{W}_{\text{even}} = \mathbf{S}^{-1} \mathbf{F}_{\cos}.$$

Since both the operators $\mathbf{S}^{-1}$ and $\mathbf{F}_{\cos}$ have inverse operators, the operator $\mathbf{W}_{\text{even}}$ has an inverse operator as well:

$$\mathbf{W}^{-1}_{\text{even}} = \mathbf{F}^{-1}_{\cos} \mathbf{S}. \tag{4.113}$$

Since all the operators $\mathbf{S}$, $\mathbf{S}^{-1}$, $\mathbf{F}_{\cos}$ and $\mathbf{F}^{-1}_{\cos}$ are bounded linear operators, so are $\mathbf{W}_{\text{even}}$ and $\mathbf{W}^{-1}_{\text{even}}$, i.e.,

$$\mathbf{W}_{\text{even}}, \mathbf{W}^{-1}_{\text{even}} \in B(L^2_{\text{even}}(R)). \tag{4.114}$$

Therefore we have

$$\begin{aligned}
f_{\text{even}}(x) &= \mathbf{W}^{-1}_{\text{even}} \, C(\omega) \\
&= \mathbf{F}^{-1}_{\cos} \mathbf{S} C(\omega) \\
&= \mathbf{F}^{-1}_{\cos} \sum_{n=1}^{\infty} A(n) \frac{1}{n} \mathbf{T}(n) C(\omega)
\end{aligned}$$

$$\begin{aligned}
&= \mathbf{F}_{\cos}^{-1}\sum_{n=1}^{\infty} A(n)\frac{1}{n}C(\frac{\omega}{n}) \\
&= \int_{-\infty}^{\infty}\cos(\omega x)\sum_{n=1}^{\infty} A(n)\frac{1}{n}C(\frac{\omega}{n})d\omega \\
&= \sum_{n=1}^{\infty} A(n)\frac{1}{n}\int_{-\infty}^{\infty}\cos(\omega x)C(\frac{\omega}{n})d\omega \\
&= \sum_{n=1}^{\infty} A(n)\int_{-\infty}^{\infty}\cos(n\omega x)C(\omega)d\omega \\
&= \int_{-\infty}^{\infty} C(\omega)\sum_{n=1}^{\infty} A(n)\cos(n\omega x)d\omega \\
&= \int_{-\infty}^{\infty} C(\omega)X(\omega x)d\omega.
\end{aligned}$$

The case $\Longleftarrow$ is similar. This completes the proof.

Similarly, we have the following two propositions.

Proposition 4.24 The operator series

$$\mathbf{R} = \sum_{n=1}^{\infty} B(n)\frac{1}{n}\mathbf{T}(n) \tag{4.115}$$

$$\mathbf{R}^{-1} = \sum_{n=1}^{\infty} B^{-1}(n)\frac{1}{n}\mathbf{T}(n) \tag{4.116}$$

converge absolutely in the sense of the norm in $B(L^2(R))$, and are mutually inverse.

Proposition 4.25 For any two functions $f_{\text{odd}}(x), D(\omega) \in L^2_{\text{odd}}(R)$, we have

$$D(\omega) = (\mathbf{W}_{\text{odd}} f_{\text{odd}})(\omega) = \int_{-\infty}^{+\infty} f_{\text{odd}}(x)\widetilde{Y}(\omega x)dx,$$

$$\Longleftrightarrow$$

$$f_{\text{odd}}(x) = (\mathbf{W}_{\text{odd}}^{-1} D)(x) = \int_{-\infty}^{+\infty} D(\omega)Y(\omega x)d\omega,$$

where $\mathbf{W}_{\text{odd}} : L^2_{\text{odd}}(R) \to L^2_{\text{odd}}(R)$ denotes the frequency transform based on odd periodic function $Y(x)$, and $\mathbf{W}_{\text{odd}}^{-1} : L^2_{\text{odd}}(R) \to L^2_{\text{odd}}(R)$ denotes its inverse.

Proposition 4.26 For any function $f(x) \in L^2(-\infty, +\infty)$, we have

$$f(x) = \int_{-\infty}^{\infty} C(\omega)X(\omega x) + D(\omega)Y(\omega x)d\omega, \tag{4.117}$$

where

$$C(\omega) = \int_{-\infty}^{\infty} f(x)\widetilde{X}(\omega x)dx, \tag{4.118}$$

$$D(\omega) = \int_{-\infty}^{\infty} f(x)\widetilde{Y}(\omega x)dx. \tag{4.119}$$

Proof. Combining Propositions 4.23 and 4.25, we obtain this one easily.

Definition. The two functions $W(x)$ and $\widetilde{W}(x)$ are defined on $(-\infty, +\infty)$ by

$$W(x) = X(x) + Y(x), \tag{4.120}$$

$$\widetilde{W}(x) = \widetilde{X}(x) + \widetilde{Y}(x). \tag{4.121}$$

Proposition 4.27 For any two functions $f(x), f'(\omega) \in L^2(-\infty, +\infty)$, we have

$$f'(\omega) = \int_{-\infty}^{+\infty} f(x)\widetilde{W}(\omega x)dx,$$

$$\Longleftrightarrow$$

$$f(x) = \int_{-\infty}^{+\infty} f'(\omega)W(\omega x)d\omega.$$

Proof. Combining Propositions 4.23 and 4.25, we obtain this one easily. By the way, this transform is a generalization of Hartley transform [8].

Definition. The two complex-valued functions $Z(x)$ and $\widetilde{Z}(x)$ are defined on $(-\infty, +\infty)$ by

$$Z(x) = X(x) + iY(x), \tag{4.122}$$

$$\widetilde{Z}(x) = \widetilde{X}(x) - i\widetilde{Y}(x). \tag{4.123}$$

Proposition 4.28 For any two functions $f(x), \overline{f}(\omega) \in L^2(-\infty, +\infty)$, we have

$$\overline{f}(\omega) = (\mathbf{W}f)(\omega) = \int_{-\infty}^{+\infty} f(x)\widetilde{Z}(\omega x)dx,$$

$$\Longleftrightarrow$$

$$f(x) = (\mathbf{W}^{-1}\overline{f})(x) = \int_{-\infty}^{+\infty} \overline{f}(\omega)Z(\omega x)d\omega.$$

$\mathbf{W} : L^2(R) \to L^2(R)$ is called the frequency transform based on the periodic functions $X(x)$ and $Y(x)$, and $\mathbf{W}^{-1} : L^2(R) \to L^2(R)$ is its inversion.

Proof. Combining Propositions 4.23 and 4.25, we obtain this one easily.

The relation between the transform based on $X(x)$ and $Y(x)$ and Fourier transform of the same function $f(x) \in L^2(R)$ is

$$\begin{aligned} \overline{f} &= C(\omega) - iD(\omega) \\ &= \mathbf{S}^{-1}a(\omega) - i\mathbf{R}^{-1}b(\omega) \\ &= \sum_{n=1}^{\infty} [\ A(n)^{-1}\frac{1}{n}a(\frac{\omega}{n}) - iB(n)^{-1}\frac{1}{n}b(\frac{\omega}{n})\ ], \end{aligned}$$

where the even function $a(\omega)$ and the odd function $b(\omega)$ are defined by $f(x)$'s Fourier transform

$$\hat{f}(\omega) = a(\omega) - ib(\omega). \tag{4.124}$$

SUMMARY

In this chapter we develop a theory for general frequency series in $L^2[-\pi, \pi]$ and a theory for general frequency transform in $L^2(R)$, which is a generalization of Fourier series and Fourier transform. Almost every common waveform is considered as an example. A number of complete frequency systems and unconditional bases of practical importance are presented. This forms a theoretical foundation for the technique of common waveform analysis.

Unsolved Problems:

1. What is the sufficient and necessary condition for a combinative frequency system $1, \{X(nx), Y(nx)\}_{n=1}^{\infty}$ to be

(1) a complete system;

(2) a basis;

(3) an unconditional basis;

in $L^2[-\pi, \pi]$?

2. What is the sufficient and necessary conditions for a function in $L^2(R)$ to be expressed as a frequency integral (see (4.117))?

They are central questions in this field.

Chapter 5

MAIN RELATIONS AND BASIC TECHNIQUES

As an easy and compressed version of this monograph, this chapter is self-contained and independent of the previous chapters. The authors suggest the reader go through this chapter first and then the others. We pay our attention to main relations between signals and common waveforms, and related basic techniques.

In addition to sine-cosine function, sawtooth wave, square wave, triangular wave and trapezoidal wave are common waveforms as well. Since Fourier analysis provides us the relation between signals and sine-cosine functions, sine-cosine function is used widely; on the other hand, since we are unaware of the relation between signals and other common waveforms, the practical applications of square wave, triangular wave, and trapezoidal wave are limited seriously. By means of a formula related to number theory, we discover the relation between signals and common waveforms and successfully generalize Fourier analysis to common waveform analysis. Associated with Fourier series, Fourier transform, and discrete Fourier transform, we introduce common waveform series, common waveform transform, and discrete common waveform transform. These results make it possible to represent a signal as a superposition of common waveforms and form a theoretical foundation for the techniques of common waveform analysis. Many techniques based on sine-cosine functions can be translated into techniques based on common waveforms. The main contents of this chapter include

1. Dirichlet multiplication and a related formula;
2. Relations between sine waves and common waveforms;
3. Relations between two waveforms;
4. Common waveform series;
5. Common waveform transform;
6. Discrete common waveform transform;
7. Techniques of common waveform analysis.

5.1 Dirichlet Multiplication and a Related Formula.

Dirichlet multiplication in number theory [2] plays a basic role in common waveform analysis. In this section we have to briefly introduce this concept and present a related formula, which are a basis for other sections.

$1, 2, 3, \cdots$ are called positive integers. A function defined on the positive integers is called an arithmetical function. Here are three useful examples.

Example 1. The identity function I is defined by

$$I(n) = \begin{cases} 1 & n = 1, \\ 0 & n = 2, 3, 4, \cdots. \end{cases} \tag{5.1}$$

Example 2. The function ϵ is defined by

$$\epsilon(n) = 1 \qquad \text{for } n = 1, 2, 3, \cdots. \tag{5.2}$$

Example 3. The Möbius function μ

The Möbius function μ is defined by

$$\mu(n) = \begin{cases} 1, & \text{if } n = 1, \\ (-1)^r, & \text{if } n \text{ is the product of } r \text{ distinct primes}, \\ 0, & \text{if } n \text{ is divisible by a prime square.} \end{cases} \tag{5.3}$$

Notice that $\mu(n)$ assumes three values $1, -1$, and 0 only. The first ten values are $\mu(1) = 1,\ \mu(2) = -1,\ \mu(3) = -1,\ \mu(4) = 0,\ \mu(5) = -1,\ \mu(6) = 1,\ \mu(7) = -1,\ \mu(8) = 0,\ \mu(9) = 0,\ \mu(10) = 1.$

The Möbius function μ has a simple property that

$$\sum_{d|n} \mu(d) = \begin{cases} 1 & n = 1, \\ 0 & n = 2, 3, 4, \cdots, \end{cases} \tag{5.4}$$

where the summation runs over the positive factors of n including 1 and n, and $d \mid n$ means that d is a positive factor of n.

For example, for $n = 6$, we have

$$\sum_{d|6} \mu(d) = \mu(1) + \mu(2) + \mu(3) + \mu(6) = 0.$$

An arithmetical function f is called completely multiplicative [2] if f is not identically zero and

$$f(mn) = f(m)f(n) \tag{5.5}$$

for any two positive integers m and n. Obviously, if $f(n)$ is completely multiplicative, then $f(1) = 1$.

For example, the functions $I(n)$ and $\epsilon(n)$ are completely multiplicative. The function $f(n) = \frac{1}{n}$ is completely multiplicative as well.

For two given arithmetical functions f and g, their Dirichlet product is the arithmetical function h defined by the equation

$$h(n) = \sum_{d|n} f(d)g\left(\frac{n}{d}\right), \tag{5.6}$$

for $n = 1, 2, 3, \cdots$. We write

$$h = f * g. \tag{5.7}$$

In the notation of Dirichlet multiplication, the relation (5.4) can be written as

$$\mu * \epsilon = I. \tag{5.8}$$

For a given arithmetical function f, if there exists a function f^{-1} such that

$$f * f^{-1} = I, \tag{5.9}$$

then f^{-1} is called the Dirichlet inverse of f. If $f(1) \neq 0$, f has a unique inverse, and if $f(1) = 0$, f has not any inverse,

As an example, The Möbius function $\mu(n)$ and the function $\epsilon(n)$ are mutually inverse.

If f is completely multiplicative, then

$$f^{-1}(n) = \mu(n)f(n). \tag{5.10}$$

If not, it is somewhat difficult to find its inversion.

Theorem 270 in [9] provides us a miraculous relation:

$$G(x) = \sum_{n=1}^{\infty} g(nx) \Longleftrightarrow g(x) = \sum_{n=1}^{\infty} \mu(n)G(nx). \tag{5.11}$$

What we shall use many times is a generalization of this theorem.

Formula $G(x)$ and $g(x)$ are two functions defined on $(-\infty, \infty)$. $c(n)$ is a function defined on positive integers and $c^{-1}(n)$ is its Dirichlet inverse. We have

$$G(x) = \sum_{n=1}^{\infty} c(n)g(nx) \Longleftrightarrow g(x) = \sum_{n=1}^{\infty} c^{-1}(n)G(nx). \tag{5.12}$$

In fact, we have

$$\sum_{n=1}^{\infty} c^{-1}(n)G(nx)$$

$$
\begin{aligned}
&= \sum_{n=1}^{\infty} c^{-1}(n) \sum_{m=1}^{\infty} c(m) g(mnx) \\
&= \sum_{k=1}^{\infty} \sum_{d|k} c^{-1}(d) c(\frac{k}{d})\ g(kx) \\
&= \sum_{k=1}^{\infty} (c^{-1} * c)(k)\ g(kx) \\
&= \sum_{k=1}^{\infty} I(k) g(kx) \\
&= g(x),
\end{aligned}
$$

and vice versa.

5.2 Relations between Sine Waves and Common Waveforms

In this chapter, we suppose that $X(x)$ and $Y(x)$ are an even and odd function with period 2π, and that their Fourier series are

$$X(x) = \sum_{n=1}^{\infty} A(n) \cos nx, \tag{5.13}$$

$$Y(x) = \sum_{n=1}^{\infty} B(n) \sin nx, \tag{5.14}$$

with $A(1) \neq 0$ and $B(1) \neq 0$.

By relation (5.12), we obtain that

$$\cos x = \sum_{n=1}^{\infty} A^{-1}(n) X(nx), \tag{5.15}$$

$$\sin x = \sum_{n=1}^{\infty} B^{-1}(n) Y(nx). \tag{5.16}$$

Furthermore, if $A(n)$ and $B(n)$ are completely multiplicative then we have

$$\cos x = \sum_{n=1}^{\infty} \mu(n) A(n) X(nx), \tag{5.17}$$

$$\sin x = \sum_{n=1}^{\infty} \mu(n)B(n)Y(nx). \tag{5.18}$$

In this case, the functions $X(x)$ and $Y(x)$ are called functions **with** completely **multiplicative** **F**ourier coefficients, or, WCMFC functions. It is interesting that many common waveforms in electronics have completely multiplicative Fourier coefficients.

The following are some important examples of WCMFC functions. See figure 5.1.

1. Sawtooh wave.

The function $Y_{\text{sa}}(x)$ with period 2π is called a sawtooth wave, whose value in one period is given by

$$Y_{\text{sa}}(x) = \begin{cases} (\pi - x)/2 & 0 < x < 2\pi, \\ 0 & x = 0. \end{cases} \tag{5.19}$$

(In Chapter 4 we denote the function by Y_{Isa}.)

Its Fourier series is [3]

$$Y_{\text{sa}}(x) = \sin x + \frac{1}{2}\sin(2x) + \frac{1}{3}\sin(3x) + \cdots + \frac{1}{n}\sin(nx) + \cdots. \tag{5.20}$$

Its Fourier coefficients $B_{\text{sa}}(n) = \frac{1}{n}$ is completely multiplicative, and therefore $B_{\text{sa}}^{-1}(n) = \frac{\mu(n)}{n}$.

From (5.18), we obtain the sawtooh wave expansion of $\sin x$:

$$\sin x = Y_{\text{sa}}(x) - \frac{1}{2}Y_{\text{sa}}(2x) - \frac{1}{3}Y_{\text{sa}}(3x) + \cdots + \frac{\mu(n)}{n}Y_{\text{sa}}(nx) + \cdots. \tag{5.21}$$

2. Square wave.

The functions $X_{\text{sq}}(x) = \frac{\pi}{4}\text{sgn}(\cos x)$ and $Y_{\text{sq}}(x) = \frac{\pi}{4}\text{sgn}(\sin x)$ are called even and odd square waves. Here sgn denotes the sign function:

$$\text{sgn}(x) = \begin{cases} 1 & x > 0, \\ 0 & x = 0, \\ -1 & x < 0. \end{cases}$$

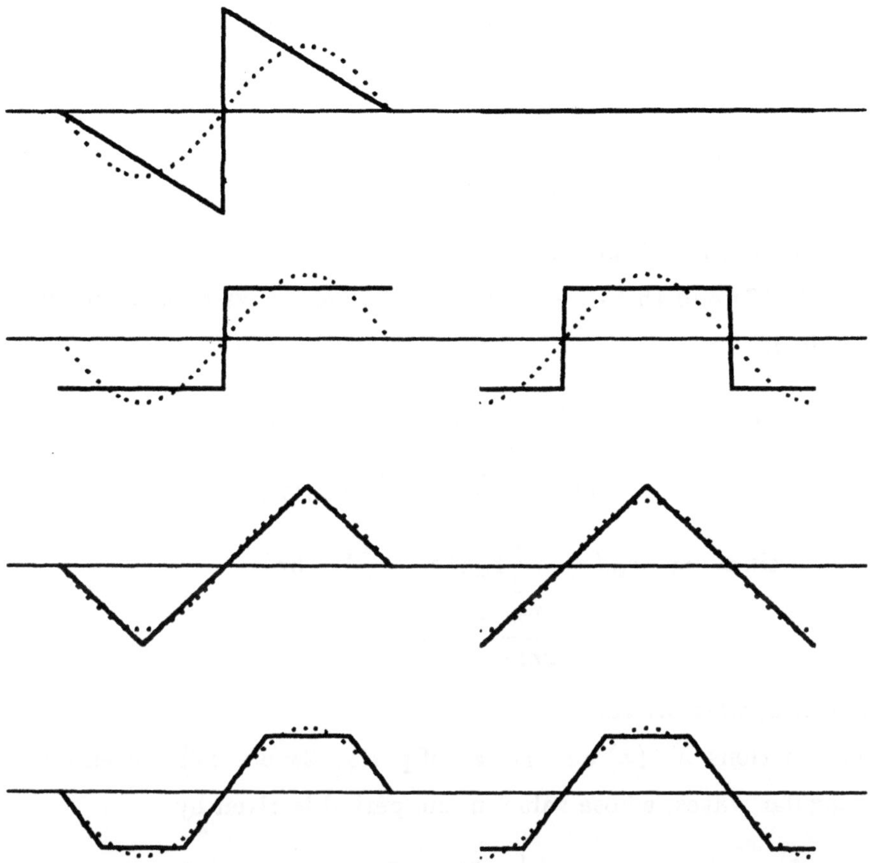

Figure 5.1: The sawtooh wave, square waves, triangular waves, and trapezoidal waves.

$$\begin{pmatrix} Y_{sa}(x) & \\ Y_{sq}(x) & X_{sq}(x) \\ Y_{tri}(x) & X_{tri}(x) \\ Y_{tra}(x) & X_{tra}(x) \end{pmatrix}$$

Their Fourier series are [3]

$$\begin{aligned} X_{sq}(x) &= \cos x - \frac{1}{3}\cos(3x) + \cdots + \frac{(-1)^m}{2m+1}\cos((2m+1)x) + \cdots \\ Y_{sq}(x) &= \sin x + \frac{1}{3}\sin(3x) + \cdots + \frac{1}{2m+1}\sin((2m+1)x) + \cdots . \end{aligned}$$

Their Fourier coeeficients

$$A_{sq}(n) = \begin{cases} (-1)^{(n-1)/2}\frac{1}{n} & n = 1, 3, 5, \cdots, \\ 0 & n = 2, 4, 6, \cdots, \end{cases} \tag{5.22}$$

$$B_{sq}(n) = \begin{cases} \frac{1}{n} & n = 1, 3, 5, \cdots, \\ 0 & n = 2, 4, 6, \cdots . \end{cases} \tag{5.23}$$

are completely multiplicative as well.

From (5.17) and (5.18), we obtain the square wave expansions of $\cos x$ and $\sin x$:

$$\begin{aligned} \cos x &= X_{sq}(x) + \frac{1}{3}X_{sq}(3x) - \frac{1}{5}X_{sq}(5x) + \cdots \\ &\quad + (-1)^m \frac{\mu(2m+1)}{2m+1} X_{sq}((2m+1)x) + \cdots, \end{aligned} \tag{5.24}$$

$$\begin{aligned} \sin x &= Y_{sq}(x) - \frac{1}{3}Y_{sq}(3x) - \frac{1}{5}Y_{sq}(5x) + \cdots \\ &\quad + \frac{\mu(2m+1)}{2m+1} Y_{sq}((2m+1)x) + \cdots . \end{aligned} \tag{5.25}$$

3. Triangular wave.

The functions $X_{tri}(x)$ and $Y_{tri}(x)$ of period 2π are called even and odd triangular waves, whose value in one period is given by

$$X_{tri}(x) = \begin{cases} \frac{\pi}{4}(x + \frac{\pi}{2}) & -\pi \le x < 0, \\ \frac{\pi}{4}(-x + \frac{\pi}{2}) & 0 \le x < \pi, \end{cases} \tag{5.26}$$

$$Y_{tri}(x) = \begin{cases} \frac{\pi}{4}x & -\frac{\pi}{2} \le x < \frac{\pi}{2}, \\ \frac{\pi^2 - \pi x}{4} & \frac{\pi}{2} \le x < \frac{3\pi}{2}. \end{cases} \tag{5.27}$$

Their Fourier series are [3]

$$X_{tri}(x) = \cos x + \frac{1}{3^2}\cos(3x) + \frac{1}{5^2}\cos(5x) + \cdots$$

$$
\begin{aligned}
& + \frac{1}{(2m+1)^2} \cos((2m+1)x) + \cdots, \\
Y_{\text{tri}}(x) &= \sin x - \frac{1}{3^2} \sin(3x) + \frac{1}{5^2} \sin(5x) + \cdots \\
& + (-1)^m \frac{1}{(2m+1)^2} \sin((2m+1)x) + \cdots.
\end{aligned}
$$

Their Fourier coefficients

$$
A_{\text{tri}}(n) = \begin{cases} \frac{1}{n^2} & n = 1, 3, 5, \cdots \\ 0 & n = 2, 4, 6, \cdots \end{cases} \tag{5.28}
$$

$$
B_{\text{tri}}(n) = \begin{cases} (-1)^{(n-1)/2} \frac{1}{n^2} & n = 1, 3, 5, \cdots \\ 0 & n = 2, 4, 6, \cdots \end{cases} \tag{5.29}
$$

are completely multiplicative as well.

From (5.17) and (5.18), we obtain the triangular wave expansions of $\cos x$ and $\sin x$:

$$
\begin{aligned}
\cos x &= X_{\text{tri}}(x) - \frac{1}{9} X_{\text{tri}}(3x) - \frac{1}{25} X_{\text{tri}}(5x) + \cdots \\
& + \frac{\mu(2m+1)}{(2m+1)^2} X_{\text{tri}}((2m+1)x) + \cdots, \qquad (5.30) \\
\sin x &= Y_{\text{tri}}(x) + \frac{1}{9} Y_{\text{tri}}(3x) - \frac{1}{25} Y_{\text{tri}}(5x) + \cdots \\
& + (-1)^m \frac{\mu(2m+1)}{(2m+1)^2} Y_{\text{tri}}((2m+1)x) + \cdots. \qquad (5.31)
\end{aligned}
$$

4. Trapezoidal wave

The functions $X_{\text{tra}}(x)$ and $Y_{\text{tra}}(x)$ with period 2π are called an even trapezoidal wave and an odd trapezoidal wave, whose value in one period is given by

$$
X_{\text{tra}}(x) = \begin{cases} -\frac{\sqrt{2}\pi^2}{16} & x \in [-\frac{5\pi}{4}, -\frac{3\pi}{4}], \\ \frac{\sqrt{2}\pi}{4}(x + \frac{\pi}{2}) & x \in [-\frac{3\pi}{4}, \frac{-\pi}{4}], \\ \frac{\sqrt{2}\pi^2}{16} & x \in [\frac{-\pi}{4}, \frac{\pi}{4}], \\ \frac{\sqrt{2}\pi}{4}(\frac{\pi}{2} - x) & x \in [\frac{\pi}{4}, \frac{3\pi}{4}], \end{cases} \tag{5.32}
$$

$$Y_{\rm tra}(x) = \begin{cases} -\frac{\sqrt{2}\pi^2}{16} & x \in [-\frac{3\pi}{4}, -\frac{\pi}{4}], \\ \frac{\sqrt{2}\pi}{4}x & x \in [-\frac{\pi}{4}, \frac{\pi}{4}], \\ \frac{\sqrt{2}\pi^2}{16} & x \in [\frac{\pi}{4}, \frac{3\pi}{4}], \\ \frac{\sqrt{2}\pi}{4}(\pi - x) & x \in [\frac{3\pi}{4}, \frac{5\pi}{4}]. \end{cases} \qquad (5.33)$$

(In Chapter 3 we denote the two functions by $X_{\rm Itra}(x), Y_{\rm Itra}(x)$.)
Their Fourier series are [3]

$$\begin{aligned} X_{\rm tra}(x) &= \cos x - \frac{1}{3^2}\cos(3x) - \frac{1}{5^2}\cos(5x) + \cdots \\ &\quad + \sqrt{2}\cos[(2m+1)\frac{\pi}{4}]\frac{1}{(2m+1)^2}\cos((2m+1)x) + \cdots \end{aligned}$$

$$\begin{aligned} Y_{\rm tra}(x) &= \sin x + \frac{1}{3^2}\sin(3x) - \frac{1}{5^2}\sin(5x) + \cdots \\ &\quad + \sqrt{2}\sin[(2m+1)\frac{\pi}{4}]\frac{1}{(2m+1)^2}\sin((2m+1)x) + \cdots. \end{aligned}$$

Their Fourier coefficients

$$A_{\rm tra}(n) = \begin{cases} \sqrt{2}\cos(\frac{n\pi}{4})\frac{1}{n^2} & n = 1, 3, 5, \cdots \\ 0 & n = 2, 4, 6, \cdots \end{cases} \qquad (5.34)$$

$$B_{\rm tra}(n) = \begin{cases} \sqrt{2}\sin(\frac{n\pi}{4})\frac{1}{n^2} & n = 1, 3, 5, \cdots \\ 0 & n = 2, 4, 6, \cdots \end{cases} \qquad (5.35)$$

are completely multiplicative as well.
From (5.17) and (5.18), we obtain the trapezoidal wave expansions of $\cos x$ and $\sin x$:

$$\begin{aligned} \cos x &= X_{\rm tra}(x) + \frac{1}{9}X_{\rm tra}(3x) + \frac{1}{25}X_{\rm tra}(5x) + \cdots \\ &\quad + \sqrt{2}\cos[(2m+1)\frac{\pi}{4}]\frac{\mu(2m+1)}{(2m+1)^2}X_{\rm tra}((2m+1)x) + \cdots \qquad (5.36) \\ \sin x &= Y_{\rm tra}(x) - \frac{1}{9}Y_{\rm tra}(3x) + \frac{1}{25}Y_{\rm tra}(5x) + \cdots \\ &\quad + \sqrt{2}\sin[(2m+1)\frac{\pi}{4}]\frac{\mu(2m+1)}{(2m+1)^2}Y_{\rm tra}((2m+1)x) + \cdots. \qquad (5.37) \end{aligned}$$

The convergence processes of the series (5.21),(5.25), (5.31) and (5.37) are shown in Figure 5.2.

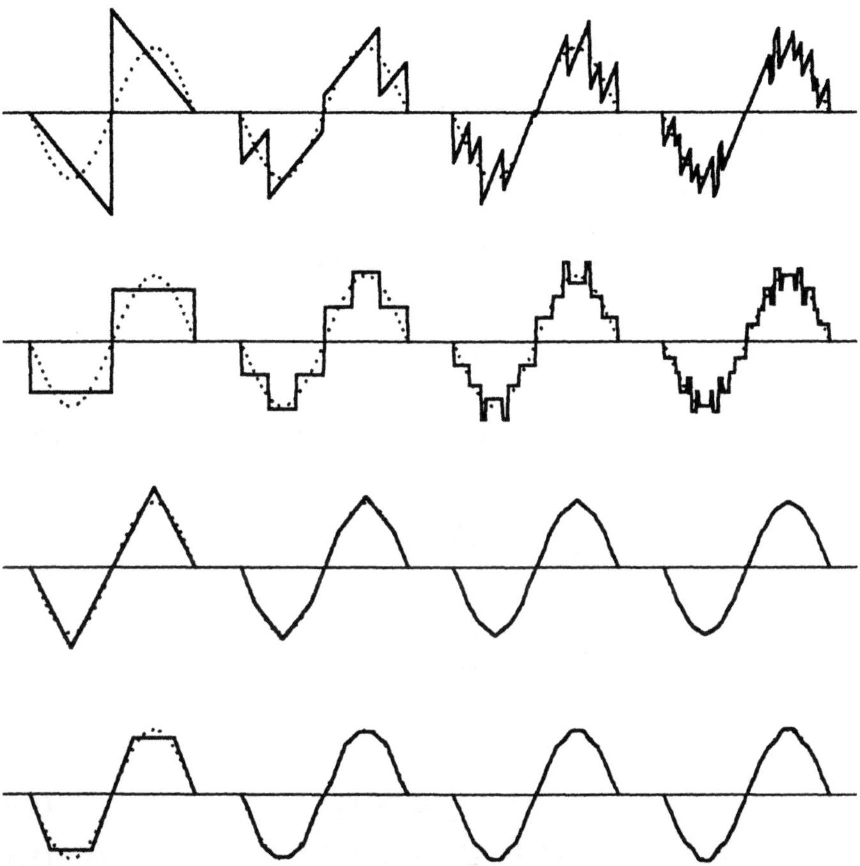

Figure 5.2: The superposition of common waveforms of different frequencies $\sum_{n=1}^{N} \mu(n)B(n)Y(nx)$ approximates $\sin x$ as the term number N increase.

(N=1,3,5,7.)

5.3 Relations between Two Waveforms

Suppose that $X_1(x)$ and $X_2(x)$ are two even functions with period 2π, and their Fourier cosine series are

$$X_1(x) = \sum_{n=1}^{\infty} A_1(n)\cos nx, \tag{5.38}$$

$$X_2(x) = \sum_{n=1}^{\infty} A_2(n)\cos nx, \tag{5.39}$$

with $A_1(1) \neq 0$ and $A_2(1) \neq 0$.

Considering (5.15), we have

$$\cos x = \sum_{n=1}^{\infty} A_1^{-1}(n)X_1(nx), \tag{5.40}$$

$$\cos x = \sum_{n=1}^{\infty} A_2^{-1}(n)X_2(nx), \tag{5.41}$$

Putting (5.40) into (5.39) and putting (5.41) into (5.38), we obtain that

$$X_1(x) = \sum_{n=1}^{\infty} C(n)X_2(nx), \tag{5.42}$$

$$X_2(x) = \sum_{n=1}^{\infty} C^{-1}(n)X_1(nx), \tag{5.43}$$

where

$$C = A_1 * A_2^{-1} \tag{5.44}$$

$$C^{-1} = A_2 * A_1^{-1}. \tag{5.45}$$

For example, the relations between the even square wave $X_{sq}(x)$ and the even triangular wave $X_{tri}(x)$ are

$$X_{sq}(x) = X_{tri}(x) - \tfrac{4}{9}X_{tri}(3x) + \tfrac{4}{25}X_{tri}(5x) + \cdots$$
$$+\frac{1}{(2m+1)^2}\sum_{d|(2m+1)}(-1)^{(d-1)/2}d\mu\left(\frac{2m+1}{d}\right)X_{tri}((2m+1)x) + \cdots, \tag{5.46}$$

and

$$X_{\rm tri}(x) = X_{\rm sq}(x) + \tfrac{4}{9}X_{\rm sq}(3x) - \tfrac{4}{25}X_{\rm sq}(5x) + \cdots$$
$$+\tfrac{1}{(2m+1)^2}\sum_{d|(2m+1)}(-1)^{(d-1)/2}d\mu(d)\ X_{\rm sq}((2m+1)x) + \cdots. \tag{5.47}$$

The convergence processes of the series (5.46) and (5.47) are shown in Figure 5.3.

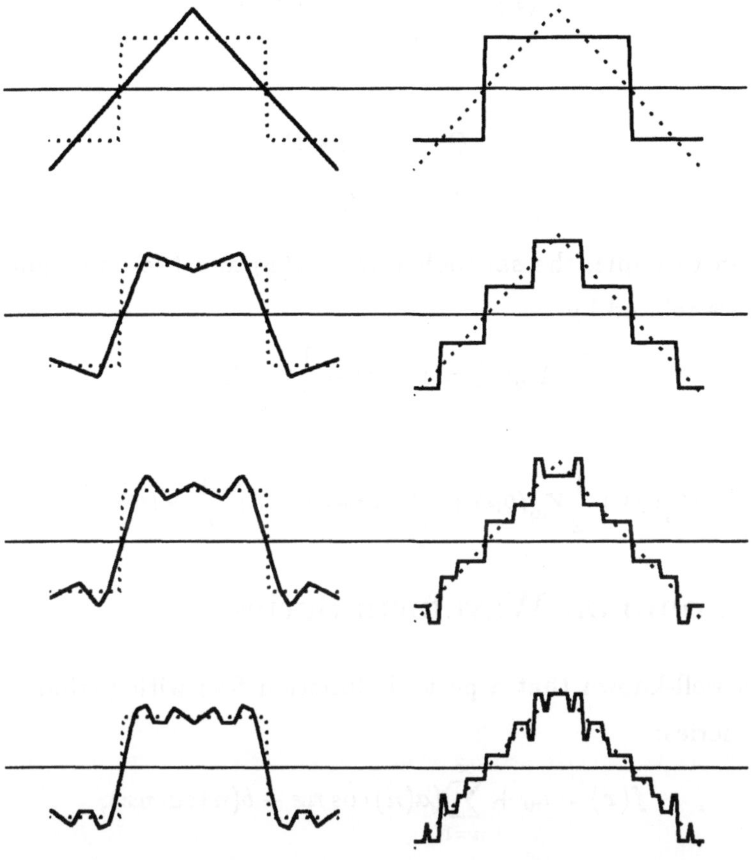

Figure 5.3: The superposition of triangular waves with different frequencies $\sum_{n=1}^{N} C_{\rm sq-tri}X_{\rm tri}(nx)$ can approximate the square wave $X_{\rm sq}(x)$, and the superposition of square waves with different frequencies $\sum_{n=1}^{N} C_{\rm tri-sq}X_{\rm sq}(nx)$ can approximate the triangular wave $X_{\rm tri}(x)$, as the term number n increases.(N=1,3,5,7.)

The relations between two odd periodic functions are similar.

Suppose that $Y_1(x)$, $Y_2(x)$ are two odd functions with period 2π, and their Fourier coefficients are $B_1(n)$ and $B_2(n)$ with $B_1(1) \neq 0$ and $B_2(1) \neq 0$. The functions $Y_1(x)$ and $Y_2(x)$ are related by

$$Y_1(x) = \sum_{n=1}^{\infty} D(n)Y_2(nx), \tag{5.48}$$

$$Y_2(x) = \sum_{n=1}^{\infty} D^{-1}(n)Y_1(nx), \tag{5.49}$$

where

$$D = B_1 * B_2^{-1} \tag{5.50}$$

$$D^{-1} = B_2 * B_1^{-1}. \tag{5.51}$$

As an example, the sawtooh wave $Y_{sa}(x)$ and the odd square wave $Y_{sq}(x)$ are related by

$$Y_{sq}(x) = Y_{sa}(x) - \frac{1}{2}Y_{sa}(2x), \tag{5.52}$$

and

$$Y_{sa}(x) = Y_{sq}(x) + \frac{1}{2}Y_{sq}(2x) + \frac{1}{4}Y_{sq}(4x) + \cdots + \frac{1}{2^m}Y(2^m x) + \cdots. \tag{5.53}$$

5.4 Common Waveform Series

It is well-known that a periodic function f(x) with period 2π has a Fourier series:

$$f(x) = a_0 + \sum_{n=1}^{\infty}(a(n)\cos nx + b(n)\sin nx), \tag{5.54}$$

where the Fourier coefficients are determined by

$$a_0 = \frac{1}{2\pi}\int_{-\pi}^{+\pi} f(x)dx,$$

$$a(n) = \frac{1}{\pi}\int_{-\pi}^{+\pi} f(x)\cos nx dx,$$

$$b(n) = \frac{1}{\pi}\int_{-\pi}^{+\pi} f(x)\sin nx dx.$$

Now let us introduce frequency series based on general periodic functions $X(x)$ and $Y(x)$.

Putting (5.15) and (5.16) into (5.54), we obtain

$$\begin{aligned} f(x) &= a_0 + \sum_{n=1}^{\infty} a(n) \sum_{l=1}^{\infty} A^{-1}(l) X(nlx) + \sum_{n=1}^{\infty} b(n) \sum_{l=1}^{\infty} B^{-1}(l) Y(nlx) \\ &= a_0 + \sum_{n=1}^{\infty} (\sum_{d\mid n} a(d) A^{-1}(\frac{n}{d})) X(nx) + \sum_{n=1}^{\infty} (\sum_{d\mid n} b(d) B^{-1}(\frac{n}{d})) Y(nx) \\ &= C_0 + \sum_{n=1}^{\infty} (C(n) X(nx) + D(n) Y(nx)), \end{aligned} \tag{5.55}$$

where the coefficients are

$$\begin{aligned} C_0 &= a_0, \\ C(n) &= \sum_{d\mid n} A^{-1}(\frac{n}{d}) a(d), \\ D(n) &= \sum_{d\mid n} B^{-1}(\frac{n}{d}) b(d), \end{aligned} \tag{5.56}$$

or in notation of Dirichlet multiplication

$$\begin{aligned} C_0 &= a_0, \\ C &= A^{-1} * a, \\ D &= B^{-1} * b. \end{aligned} \tag{5.57}$$

If we define $h_n(x)$ and $g_n(x)$ as

$$\begin{aligned} h_n(x) &= \frac{1}{\pi} \sum_{d\mid n} A^{-1}(\frac{n}{d}) \cos dx, \\ g_n(x) &= \frac{1}{\pi} \sum_{d\mid n} B^{-1}(\frac{n}{d}) \sin dx, \end{aligned}$$

then the coefficient formula (5.56) can be rewritten as

$$\begin{aligned} C_0 &= \frac{1}{2\pi} \int_{-\pi}^{+\pi} f(x) dx, \\ C(n) &= \int_{-\pi}^{+\pi} f(x) h_n(x) dx, \\ D(n) &= \int_{-\pi}^{+\pi} f(x) g_n(x) dx. \end{aligned} \tag{5.58}$$

The series (5.55) is called frequency series based on general periodic functions $X(x)$ and $Y(x)$. When $X(x)$ and $Y(x)$ are common waveforms, the series (5.55) is called common waveform series. The functions $h_n(x)$ and $g_n(x)$ are called the biorthogonal functions related to the periodic functions $X(x)$ and $Y(x)$. By Eq. (5.58), a signal can be expressed as a frequency series based on an arbitrary waveform, such as square wave, triangular wave, trapezoidal wave and so on.

Example 1. Square wave series

The square wave expansion of a function $f(x)$ is

$$f(x) = C_0 + \sum_{n=1}^{\infty}(C(n)X_{\text{sq}}(nx) + D(n)Y_{\text{sq}}(nx)), \tag{5.59}$$

where the coefficients can be determined by

$$\begin{aligned} C_0 &= \frac{1}{2\pi}\int_{-\pi}^{+\pi} f(x)dx, \\ C(n) &= \int_{-\pi}^{+\pi} f(x)h_n^{\text{sq}}(x)dx, \\ D(n) &= \int_{-\pi}^{+\pi} f(x)g_n^{\text{sq}}(x)dx. \end{aligned}$$

By the definition, the biorthogonal functions $h_n^{\text{sq}}(x)$ and $g_n^{\text{sq}}(x)$ are

$$\begin{aligned} h_n^{\text{sq}}(x) &= \frac{1}{\pi}\sum_{d|n}\mu(\frac{n}{d})A_{\text{sq}}(\frac{n}{d})\cos dx, \\ g_n^{\text{sq}}(x) &= \frac{1}{\pi}\sum_{d|n}\mu(\frac{n}{d})B_{\text{sq}}(\frac{n}{d})\sin dx. \end{aligned}$$

The first few biorthogonal functions of square wave are

$$\begin{aligned} h_1^{\text{sq}}(x) &= \frac{1}{\pi}\cos x, \\ h_2^{\text{sq}}(x) &= \frac{1}{\pi}\cos 2x, \\ h_3^{\text{sq}}(x) &= \frac{1}{\pi}(\cos 3x + \frac{1}{3}\cos x), \end{aligned}$$

and

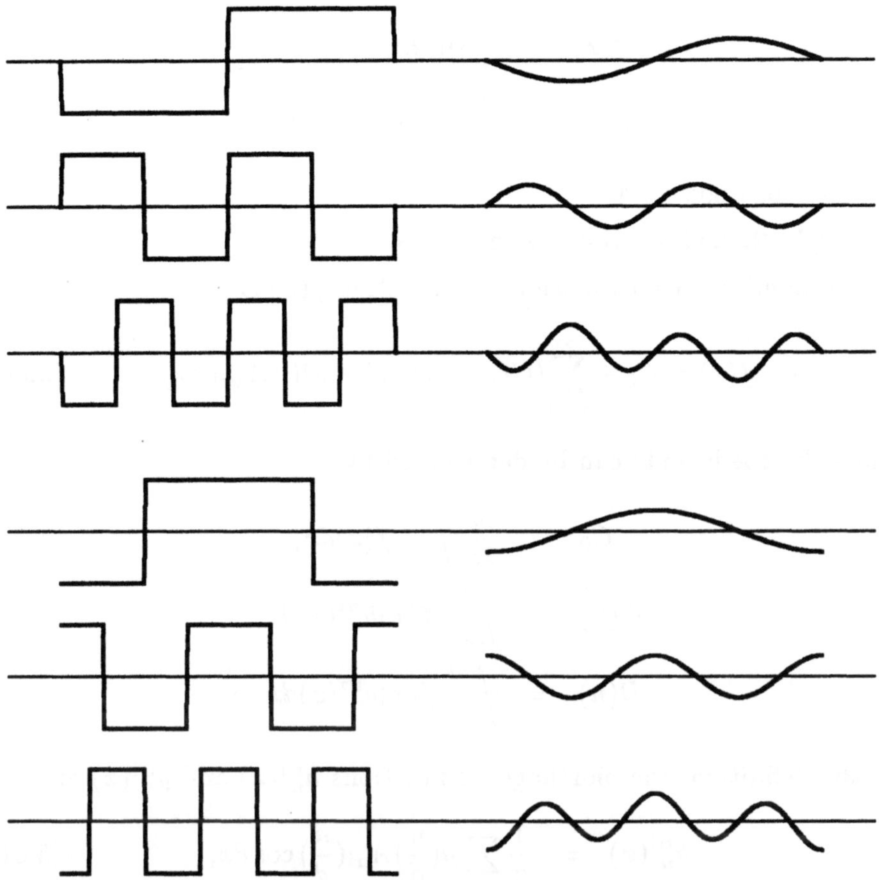

Figure 5.4: Square waves and their biorthogonal functions.

$$\begin{pmatrix} X_{sq}(x) & h_1^{sq}(x) \\ X_{sq}(2x) & h_2^{sq}(x) \\ X_{sq}(3x) & h_3^{sq}(x) \\ Y_{sq}(x) & g_1^{sq}(x) \\ Y_{sq}(2x) & g_2^{sq}(x) \\ Y_{sq}(3x) & g_3^{sq}(x) \end{pmatrix}$$

$$\begin{aligned}
g_1^{\mathrm{sq}}(x) &= \frac{1}{\pi}\sin x,\\
g_2^{\mathrm{sq}}(x) &= \frac{1}{\pi}\sin 2x,\\
g_3^{\mathrm{sq}}(x) &= \frac{1}{\pi}(\sin 3x - \frac{1}{3}\sin x),
\end{aligned}$$

as shown in Figure 5.4.

Example 2. Triangular wave series.

The triangular wave expansion of a function $f(x)$ is

$$f(x) = C_0 + \sum_{n=1}^{\infty}(C(n)X_{\mathrm{tri}}(nx) + D(n)Y_{\mathrm{tri}}(nx)), \tag{5.60}$$

where the coefficients can be determined by

$$\begin{aligned}
C_0 &= \frac{1}{2\pi}\int_{-\pi}^{+\pi} f(x)dx,\\
C(n) &= \int_{-\pi}^{+\pi} f(x)h_n^{\mathrm{tri}}(x)dx,\\
D(n) &= \int_{-\pi}^{+\pi} f(x)g_n^{\mathrm{tri}}(x)dx.
\end{aligned}$$

By the definition, the biorthogonal functions $h_n^{\mathrm{tri}}(x)$ and $g_n^{\mathrm{tri}}(x)$ are

$$h_n^{\mathrm{tri}}(x) = \frac{1}{\pi}\sum_{d|n}\mu(\frac{n}{d})A_{\mathrm{tri}}(\frac{n}{d})\cos dx, \tag{5.61}$$

$$g_n^{\mathrm{tri}}(x) = \frac{1}{\pi}\sum_{d|n}\mu(\frac{n}{d})B_{\mathrm{tri}}(\frac{n}{d})\sin dx. \tag{5.62}$$

The first few biorthogonal functions of triangular wave are

$$\begin{aligned}
h_1^{\mathrm{tri}}(x) &= \frac{1}{\pi}\cos x,\\
h_2^{\mathrm{tri}}(x) &= \frac{1}{\pi}\cos 2x,\\
h_3^{\mathrm{tri}}(x) &= \frac{1}{\pi}(\cos 3x - \frac{1}{9}\cos x),
\end{aligned}$$

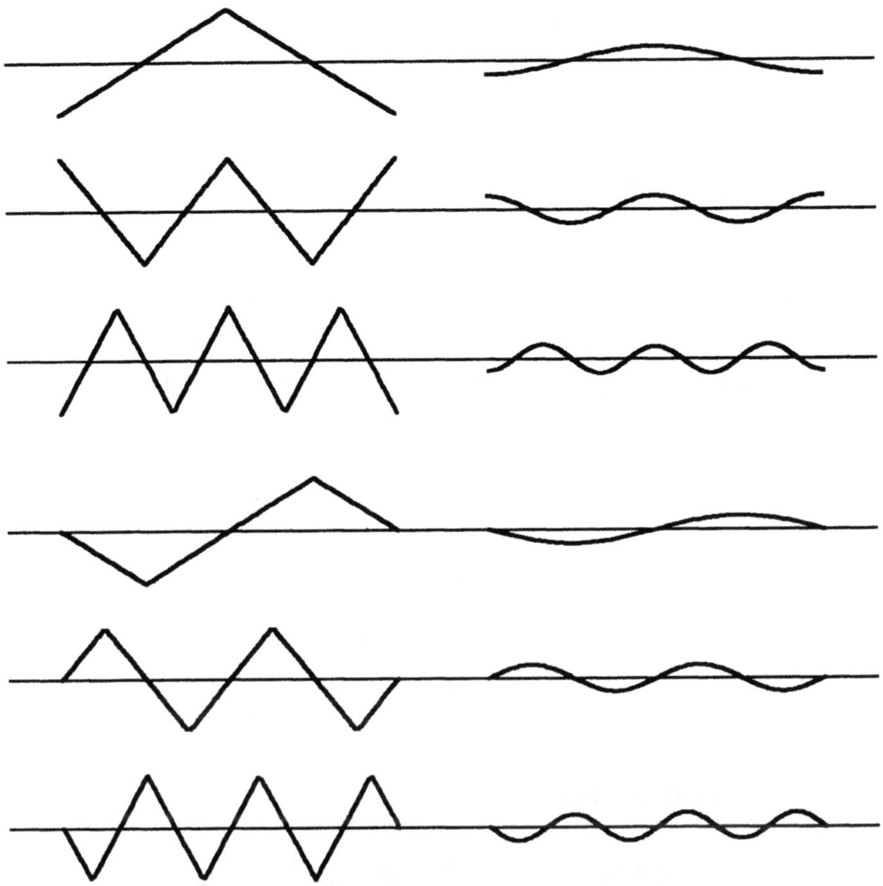

Figure 5.5: Triangular waves and their biorthogonal functions.

$$\begin{pmatrix} X_{\mathrm{tri}}(x) & h_1^{\mathrm{tri}}(x) \\ X_{\mathrm{tri}}(2x) & h_2^{\mathrm{tri}}(x) \\ X_{\mathrm{tri}}(3x) & h_3^{\mathrm{tri}}(x) \\ Y_{\mathrm{tri}}(x) & g_1^{\mathrm{tri}}(x) \\ Y_{\mathrm{tri}}(2x) & g_2^{\mathrm{tri}}(x) \\ Y_{\mathrm{tri}}(3x) & g_3^{\mathrm{tri}}(x) \end{pmatrix}$$

and

$$\begin{aligned} g_1^{\text{tri}}(x) &= \frac{1}{\pi}\sin x, \\ g_2^{\text{tri}}(x) &= \frac{1}{\pi}\sin 2x, \\ g_3^{\text{tri}}(x) &= \frac{1}{\pi}(\sin 3x + \frac{1}{9}\sin x), \end{aligned}$$

as shown in Figure 5.5.

Example 3. Trapezoidal wave series.

The trapezoidal wave expansion of a function $f(x)$ is

$$f(x) = C_0 + \sum_{n=1}^{\infty}(C(n)X_{\text{tra}}(nx) + D(n)Y_{\text{tra}}(nx)), \tag{5.63}$$

where the coefficients can be determined by

$$\begin{aligned} C_0 &= \frac{1}{2\pi}\int_{-\pi}^{+\pi} f(x)dx, \\ C(n) &= \int_{-\pi}^{+\pi} f(x)h_n^{\text{tra}}(x)dx, \\ D(n) &= \int_{-\pi}^{+\pi} f(x)g_n^{\text{tra}}(x)dx. \end{aligned}$$

The biorthogonal functions are

$$\begin{aligned} h_n^{\text{tra}}(x) &= \frac{1}{\pi}\sum_{d|n}\mu(\frac{n}{d})A_{\text{tra}}(\frac{n}{d})\cos dx, \\ g_n^{\text{tra}}(x) &= \frac{1}{\pi}\sum_{d|n}\mu(\frac{n}{d})B_{\text{tra}}(\frac{n}{d})\sin dx. \end{aligned}$$

The first few biorthogonal functions of trapezoidal wave are

$$\begin{aligned} h_1^{\text{tra}}(x) &= \frac{1}{\pi}\cos x, \\ h_2^{\text{tra}}(x) &= \frac{1}{\pi}\cos 2x, \\ h_3^{\text{tra}}(x) &= \frac{1}{\pi}(\cos 3x + \frac{1}{9}\cos x), \end{aligned}$$

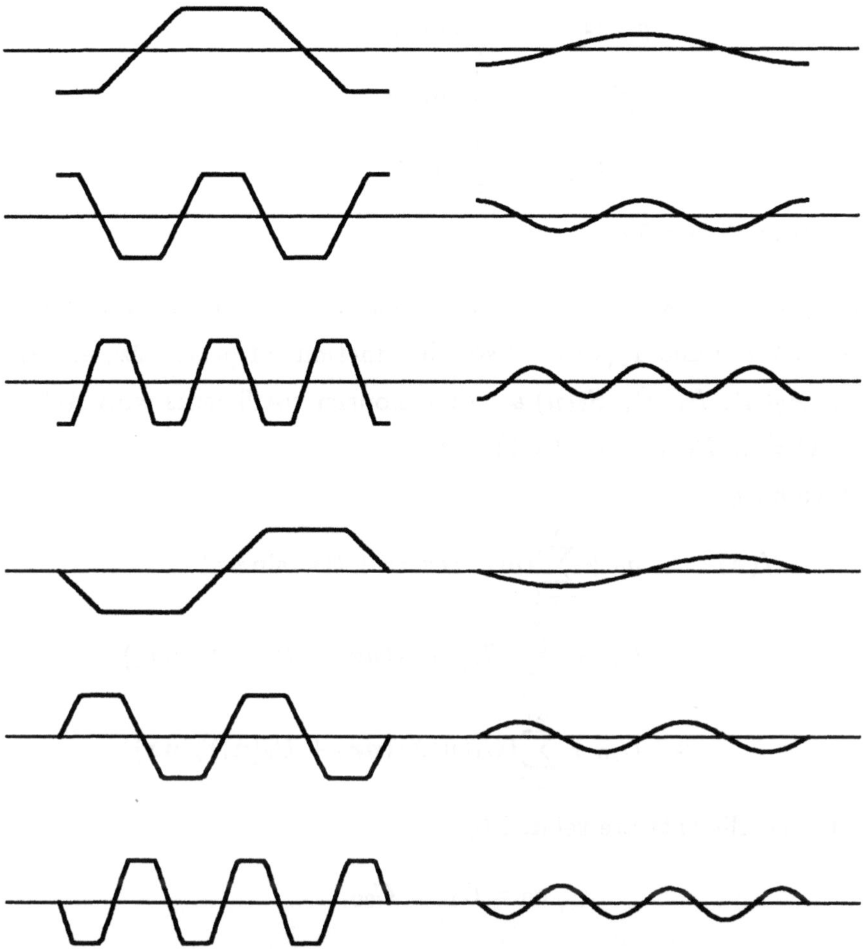

Figure 5.6: Trapezoidal waves and their biorthogonal functions.

$$\begin{pmatrix} X_{\text{tra}}(x) & h_1^{\text{tra}}(x) \\ X_{\text{tra}}(2x) & h_2^{\text{tra}}(x) \\ X_{\text{tra}}(3x) & h_3^{\text{tra}}(x) \\ Y_{\text{tra}}(x) & g_1^{\text{tra}}(x) \\ Y_{\text{tra}}(2x) & g_2^{\text{tra}}(x) \\ Y_{\text{tra}}(3x) & g_3^{\text{tra}}(x) \end{pmatrix}$$

and

$$\begin{aligned} g_1^{\text{tra}}(x) &= \frac{1}{\pi}\sin x, \\ g_2^{\text{tra}}(x) &= \frac{1}{\pi}\sin 2x, \\ g_3^{\text{tra}}(x) &= \frac{1}{\pi}(\sin 3x - \frac{1}{9}\sin x), \end{aligned}$$

as shown in Figure 5.6.

Suppose that $X_1(x)$ and $X_2(x)$ are two even functions of period 2π, and that $Y_1(x)$ and $Y_2(x)$ are two odd functions of period 2π, and that $A_1(n)$, $A_2(n)$, $B_1(n)$, $B_2(n)$ are their Fourier coefficients with $A_1(1) \neq 0$, $A_2(1) \neq 0$, $B_1(1) \neq 0$, $B_2(1) \neq 0$.

If we have

$$\begin{aligned} f(x) &= a_0 + \sum_{n=1}^{\infty}(a(n)\cos nx + b(n)\sin nx) \\ &= C_{10} + \sum_{n=1}^{\infty}(C_1(n)X_1(nx) + D_1(n)Y_1(nx)) \\ &= C_{20} + \sum_{n=1}^{\infty}(C_1(n)X_2(nx) + D_2(n)Y_2(nx)), \end{aligned}$$

then the coefficients are related by

$$a_0 = C_{10} = C_{20},$$

$$\begin{aligned} a &= C_1 * A_1 = C_2 * A_2, \\ b &= D_1 * B_1 = D_2 * B_2, \end{aligned}$$

$$C_1 = [A_2 * A_1^{-1}] \ * \ C_2, \tag{5.64}$$

$$D_1 = [B_2 * B_1^{-1}] \ * \ D_2. \tag{5.65}$$

For example, if a function $f(x)$ has a square wave series and a triangular wave series:

$$f(x) = C_0 + \sum_{n=1}^{\infty}(C_{\text{sq}}(n)X_{\text{sq}}(nx) + D_{\text{sq}}(n)Y_{\text{sq}}(nx))$$

$$= C_0 + \sum_{n=1}^{\infty}(C_{\text{tri}}(n)X_{\text{tri}}(nx) + D_{\text{tri}}(n)Y_{\text{tri}}(nx)),$$

then the coefficients are related by

$$C_{\text{sq}} = [A_{\text{tri}} * A_{\text{sq}}^{-1}] * C_{\text{tri}}, \tag{5.66}$$

$$D\text{sq} = [B_{\text{tri}} * B_{\text{sq}}^{-1}] * D_{\text{tri}}. \tag{5.67}$$

5.5 Common Waveform Transform

It is well known that a function $f(x)$ and its Fourier transform $\hat{f}(\omega)$ are related by

$$\hat{f}(\omega) = \frac{1}{2\pi}\int_{-\infty}^{+\infty} f(x)e^{-i\omega x}dx, \tag{5.68}$$

$$\Longleftrightarrow$$

$$f(x) = \int_{-\infty}^{+\infty} \hat{f}(\omega)e^{i\omega x}d\omega. \tag{5.69}$$

For an even function $f_{\text{even}}(x)$, we have

$$a(\omega) = \frac{1}{2\pi}\int_{-\infty}^{+\infty} f_{\text{even}}(x)\cos(\omega x)dx, \tag{5.70}$$

$$\Longleftrightarrow$$

$$f_{\text{even}}(x) = \int_{-\infty}^{+\infty} a(\omega)\cos(\omega x)d\omega, \tag{5.71}$$

where $a(\omega)$ is called the cosine transform of the function $f_{\text{even}}(x)$.

Similarly, for an odd function there exists the sine transform.

Now let us introduce the frequency transform based on the general periodic functions $X(x)$ and $Y(x)$.

The functions $\widetilde{X}(x)$ and $\widetilde{Y}(x)$ are called the dual functions of $X(x)$ and $Y(x)$, respectively, which are defined on $(-\infty, \infty)$ by

$$\widetilde{X}(x) = \frac{1}{2\pi}\sum_{n=1}^{\infty} A^{-1}(n)\frac{1}{n}\cos(\frac{x}{n}), \tag{5.72}$$

$$\widetilde{Y}(x) = \frac{1}{2\pi}\sum_{n=1}^{\infty} B^{-1}(n)\frac{1}{n}\sin(\frac{x}{n}). \tag{5.73}$$

For an even function $f_{\text{even}}(x)$, if

$$C(\omega) = \int_{-\infty}^{+\infty} f_{\text{even}}(x)\widetilde{X}(\omega x)dx, \tag{5.74}$$

then

$$f_{\text{even}}(x) = \int_{-\infty}^{+\infty} C(\omega)X(\omega x)d\omega. \tag{5.75}$$

In fact, we have

$$\begin{aligned}
C(\omega) &= \int_{-\infty}^{+\infty} f_{\text{even}}(x)\widetilde{X}(\omega x)dx \\
&= \frac{1}{2\pi}\int_{-\infty}^{+\infty} f_{\text{even}}(x)\sum_{n=1}^{\infty} A^{-1}(n)\frac{1}{n}\cos(\frac{x\omega}{n})dx \\
&= \frac{1}{2\pi}\sum_{n=1}^{\infty} A^{-1}(n)\frac{1}{n}\int_{-\infty}^{+\infty} f_{\text{even}}(x)\cos(\frac{x\omega}{n})dx \\
&= \frac{1}{2\pi}\sum_{n=1}^{\infty} A^{-1}(n)\int_{-\infty}^{+\infty} f_{\text{even}}(nx)\cos(x\omega)dx \\
&= \frac{1}{2\pi}\int_{-\infty}^{+\infty}\sum_{n=1}^{\infty} A^{-1}(n)f_{\text{even}}(nx)\cos(x\omega)dx.
\end{aligned}$$

By formulas of cosine transform, we have

$$\sum_{n=1}^{\infty} A^{-1}(n)f_{\text{even}}(nx) = \int_{-\infty}^{+\infty} C(\omega)\cos(x\omega)d\omega. \tag{5.76}$$

By relation (5.12), we obtain that

$$\begin{aligned}
f_{\text{even}}(x) &= \sum_{n=1}^{\infty} A(n)\int_{-\infty}^{+\infty} C(\omega)\cos(nx\omega)d\omega \\
&= \int_{-\infty}^{+\infty} C(\omega)\sum_{n=1}^{\infty} A(n)\cos(nx\omega)d\omega \\
&= \int_{-\infty}^{+\infty} C(\omega)X(\omega x)d\omega.
\end{aligned}$$

This is the very conclusion.

For the same reason, for an odd function $f_{\text{odd}}(x)$, if

$$D(\omega) = \int_{-\infty}^{+\infty} f_{\text{odd}}(x)\widetilde{Y}(\omega x)dx, \tag{5.77}$$

then

$$f_{\mathrm{odd}}(x) = \int_{-\infty}^{+\infty} D(\omega)Y(\omega x)d\omega. \tag{5.78}$$

Generally, for a function $f(x)$ if

$$C(\omega) = \int_{-\infty}^{\infty} f(x)\tilde{X}(\omega x)dx, \tag{5.79}$$

$$D(\omega) = \int_{-\infty}^{\infty} f(x)\tilde{Y}(\omega x)dx, \tag{5.80}$$

then we have

$$f(x) = \int_{-\infty}^{\infty} C(\omega)X(\omega x) + D(\omega)Y(\omega x)d\omega. \tag{5.81}$$

This is called frequency transform based on the periodic functions $X(x)$ and $Y(x)$. When $X(x)$ and $Y(x)$ are common waveforms in electronics such as square wave, triangular wave, trapezoidal wave and so on, it is called common waveform transform.

The frequency transform based on $X(x)$ and $Y(x)$ and Fourier transform of the same function $f(x)$ are related by

$$C(\omega) = \sum_{n=1}^{\infty} A^{-1}(n)\frac{1}{n}a(\frac{\omega}{n}), \tag{5.82}$$

$$D(\omega) = \sum_{n=1}^{\infty} B^{-1}(n)\frac{1}{n}b(\frac{\omega}{n}), \tag{5.83}$$

where the even function $a(\omega)$ and the odd function $b(\omega)$ are defined by $f(x)$'s Fourier transform

$$\hat{f}(\omega) = a(\omega) - ib(\omega). \tag{5.84}$$

The following are several examples of practical importance.

Example 1. Square wave transform.

For a function $f(x)$, we have

$$f(x) = \int_{-\infty}^{\infty} C(\omega)X_{\mathrm{sq}}(\omega x) + D(\omega)Y_{\mathrm{sq}}(\omega x)d\omega, \tag{5.85}$$

where

$$\begin{aligned} C(\omega) &= \int_{-\infty}^{\infty} f(x)\widetilde{X}_{\mathrm{sq}}(\omega x)dx, && (5.86)\\ D(\omega) &= \int_{-\infty}^{\infty} f(x)\widetilde{Y}_{\mathrm{sq}}(\omega x)dx. && (5.87)\end{aligned}$$

By the definition, the dual functions of square waves are

$$\begin{aligned} \widetilde{X}_{\mathrm{sq}}(x) &= \frac{1}{2\pi}\sum_{n=1}^{\infty} A_{\mathrm{sq}}(n)\frac{\mu(n)}{n}\cos(\frac{x}{n}), && (5.88)\\ \widetilde{Y}_{\mathrm{sq}}(x) &= \frac{1}{2\pi}\sum_{n=1}^{\infty} B_{\mathrm{sq}}(n)\frac{\mu(n)}{n}\sin(\frac{x}{n}), && (5.89)\end{aligned}$$

as shown in Figure 5.7.

Example 2. Triangular wave transform.

For a function $f(x)$, we have

$$f(x) = \int_{-\infty}^{\infty} C(\omega)X_{\mathrm{tri}}(\omega x) + D(\omega)Y_{\mathrm{tri}}(\omega x)d\omega, \qquad (5.90)$$

where

$$\begin{aligned} C(\omega) &= \int_{-\infty}^{\infty} f(x)\widetilde{X}_{\mathrm{tri}}(\omega x)dx, && (5.91)\\ D(\omega) &= \int_{-\infty}^{\infty} f(x)\widetilde{Y}_{\mathrm{tri}}(\omega x)dx. && (5.92)\end{aligned}$$

The dual functions of triangular waves are

$$\begin{aligned} \widetilde{X}_{\mathrm{tri}}(x) &= \frac{1}{2\pi}\sum_{n=1}^{\infty} A_{\mathrm{tri}}(n)\frac{\mu(n)}{n}\cos(\frac{x}{n}), && (5.93)\\ \widetilde{Y}_{\mathrm{tri}}(x) &= \frac{1}{2\pi}\sum_{n=1}^{\infty} B_{\mathrm{tri}}(n)\frac{\mu(n)}{n}\sin(\frac{x}{n}), && (5.94)\end{aligned}$$

as shown in Figure 5.8.

Example 3. Trapezoidal wave transform.

For a function $f(x)$, we have

$$f(x) = \int_{-\infty}^{\infty} C(\omega)X_{\mathrm{tra}}(\omega x) + D(\omega)Y_{\mathrm{tra}}(\omega x)d\omega, \qquad (5.95)$$

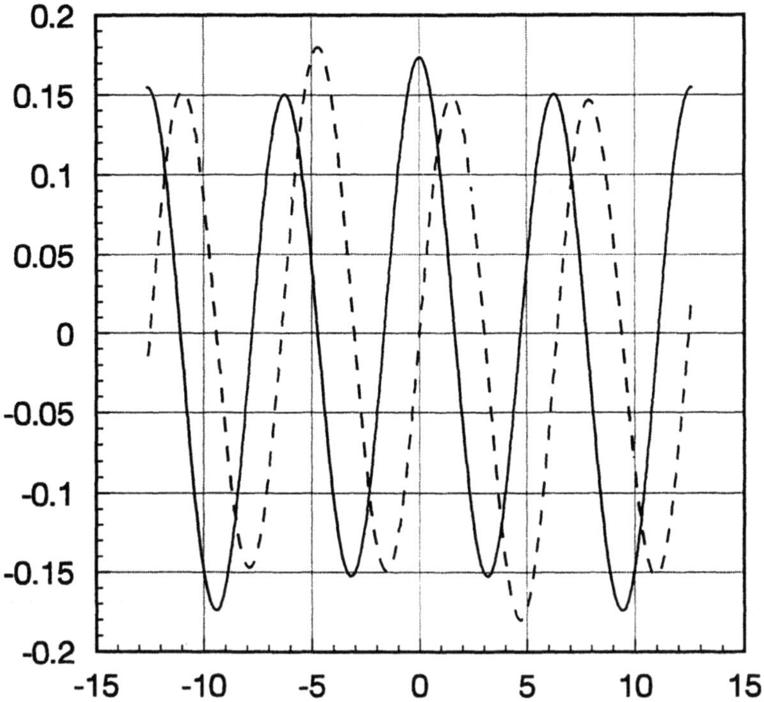

Figure 5.7: The dual functions $\widetilde{X}_{\mathrm{sq}}(x)$ and $\widetilde{Y}_{\mathrm{sq}}(x)$ (- -) of square waves.

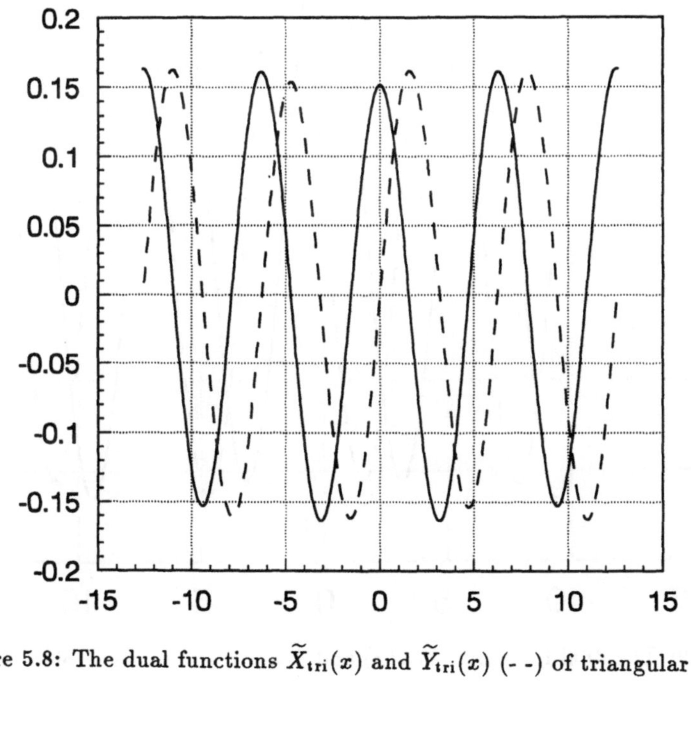

Figure 5.8: The dual functions $\widetilde{X}_{\mathrm{tri}}(x)$ and $\widetilde{Y}_{\mathrm{tri}}(x)$ (- -) of triangular waves.

where

$$C(\omega) = \int_{-\infty}^{\infty} f(x)\widetilde{X}_{\mathrm{tra}}(\omega x)dx, \tag{5.96}$$

$$D(\omega) = \int_{-\infty}^{\infty} f(x)\widetilde{Y}_{\mathrm{tra}}(\omega x)dx. \tag{5.97}$$

The dual functions of trapezoidal waves are

$$\widetilde{X}_{\mathrm{tra}}(x) = \frac{1}{2\pi}\sum_{n=1}^{\infty} A_{\mathrm{tra}}(n)\frac{\mu(n)}{n}\cos(\frac{x}{n}), \tag{5.98}$$

$$\widetilde{Y}_{\mathrm{tra}}(x) = \frac{1}{2\pi}\sum_{n=1}^{\infty} B_{\mathrm{tra}}(n)\frac{\mu(n)}{n}\sin(\frac{x}{n}), \tag{5.99}$$

as shown in Figure 5.9.

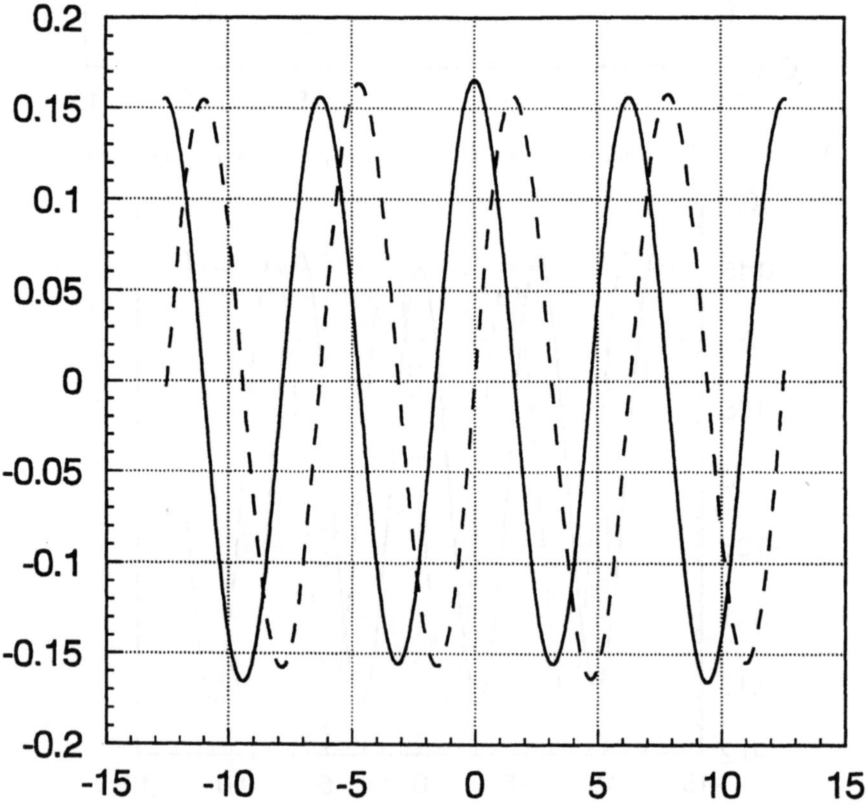

Figure 5.9: The dual functions $\widetilde{X}_{\mathrm{tra}}(x)$ and $\widetilde{Y}_{\mathrm{tra}}(x)$ (- -) of trapezoidal waves.

5.6 Discrete Transform for Common Waveform

In this section, R^N denotes the N-dimensional Euclidean space, where $N = 2m + 1$ is an odd integer. $<,>_N$ denotes the inner product in the space. $f(x)$ is a function with period 2π, but it is known only at the N points $x_k = \frac{k}{N}2\pi$, where $k = 0, 1, 2, \cdots, N-1$. For convenience we introduce the notation

$$(f(x))_N = \begin{pmatrix} f(x_0) \\ f(x_1) \\ \vdots \\ f(x_{N-1}) \end{pmatrix}, \tag{5.100}$$

which denotes a column vector consisting of N values of $f(x)$.

According to discrete Fourier transform [3], we have

$$(f(x))_N = a(0)(1)_N + \sum_{n=1}^{m} a(n)(\cos nx)_N + b(n)(\sin nx)_N, \tag{5.101}$$

where

$$\begin{aligned} a(0) &= \frac{1}{N} < 1, f(x) >_N = \frac{1}{N}\sum_{k=0}^{N-1} f(x_k), \\ a(n) &= \frac{2}{N} < \cos nx, f(x) >_N = \frac{2}{N}\sum_{k=0}^{N-1} f(x_k)\cos nx_k, \\ b(n) &= \frac{2}{N} < \sin nx, f(x) >_N = \frac{2}{N}\sum_{k=0}^{N-1} f(x_k)\sin nx_k. \end{aligned}$$

Suppose that $X(x)$ and $Y(x)$ are even and odd 2π-periodic function. Now our aim is to determine coefficients $C(0), C(n), D(n)$ such that

$$(f(x))_N = C(0)(1)_N + \sum_{n=1}^{m} C(n)(X(nx))_N + D(n)(Y(nx))_N. \tag{5.102}$$

From (5.15) and (5.16) it follows that

$$\cos ix = \sum_{n=1}^{\infty} A^{-1}(n)X(inx), \tag{5.103}$$

$$\sin ix = \sum_{n=1}^{\infty} B^{-1}(n) Y(inx), \tag{5.104}$$

for $i = 1, 2, 3, \cdots, m$. Since

$$\begin{aligned} (X[(n+N)x])_N &= (X(nx))_N, \\ (Y[(n+N)x])_N &= (Y(nx))_N, \end{aligned}$$

it follows that

$$\begin{aligned} (\cos ix)_N &= A_i^{-1}(0) X(0)\,(1)_N + \sum_{n=1}^{m} A_i^{-1}(n)(X(nx))_N, \\ (\sin ix)_N &= \sum_{n=1}^{m} B_i^{-1}(n)(Y(nx))_N, \end{aligned}$$

where

$$\begin{aligned} A_i^{-1}(0) &= \sum_{il\equiv 0} A^{-1}(l), \\ A_i^{-1}(n) &= \sum_{il\equiv n} A^{-1}(l) + \sum_{il\equiv -n} A^{-1}(l), \\ B_i^{-1}(n) &= \sum_{il\equiv n} B^{-1}(l) - \sum_{il\equiv -n} B^{-1}(l). \end{aligned} \tag{5.105}$$

Therefore we have the following relation:

$$\left((1)_N, (\cos x)_N, \cdots, (\cos mx)_N, (\sin x)_N, \cdots, (\sin mx)_N \right)$$

$$= \left((1)_N, (X(x))_N, \cdots, (X(mx))_N, (Y(x))_N, \cdots, (Y(mx))_N \right) P^{-1},$$

where

$$P^{-1} = \begin{pmatrix} 1 & A_1^{-1}(0)X(0) & \cdots & A_m^{-1}(0)X(0) & & & \\ 0 & A_1^{-1}(1) & \cdots & A_m^{-1}(1) & & 0 & \\ \vdots & \vdots & \vdots & \vdots & & & \\ 0 & A_1^{-1}(m) & \cdots & A_m^{-1}(m) & & & \\ & & & & B_1^{-1}(1) & \cdots & B_m^{-1}(1) \\ & & 0 & & \vdots & \vdots & \vdots \\ & & & & B_1^{-1}(m) & \cdots & B_m^{-1}(1) \end{pmatrix}.$$

Therefore the two group of coefficients are related by

$$\begin{pmatrix} C(0) \\ C(1) \\ \vdots \\ C(m) \\ D(1) \\ \vdots \\ D(m) \end{pmatrix} = P^{-1} \begin{pmatrix} a(0) \\ a(1) \\ \vdots \\ a(m) \\ b(1) \\ \vdots \\ b(m) \end{pmatrix}. \tag{5.106}$$

If we define

$$\begin{aligned} u_0(x) &= \frac{1}{N} + \frac{2}{N} X(0) \sum_{n=1}^{m} A_n^{-1}(0) \cos nx, \\ u_i(x) &= \frac{2}{N} \sum_{n=1}^{m} A_n^{-1}(i) \cos nx, \\ v_i(x) &= \frac{2}{N} \sum_{n=1}^{m} B_n^{-1}(i) \sin nx, \end{aligned} \tag{5.107}$$

then the coefficients in (5.102) can be calculated by

$$\begin{aligned} C(0) &= \; < u_0(x), f(x) >_N = \sum_{k=0}^{N-1} u_0(x_k) f(x_k), \\ C(i) &= \; < u_i(x), f(x) >_N = \sum_{k=0}^{N-1} u_i(x_k) f(x_k), \\ D(i) &= \; < v_i(x), f(x) >_N = \sum_{k=0}^{N-1} v_i(x_k) f(x_k), \end{aligned} \tag{5.108}$$

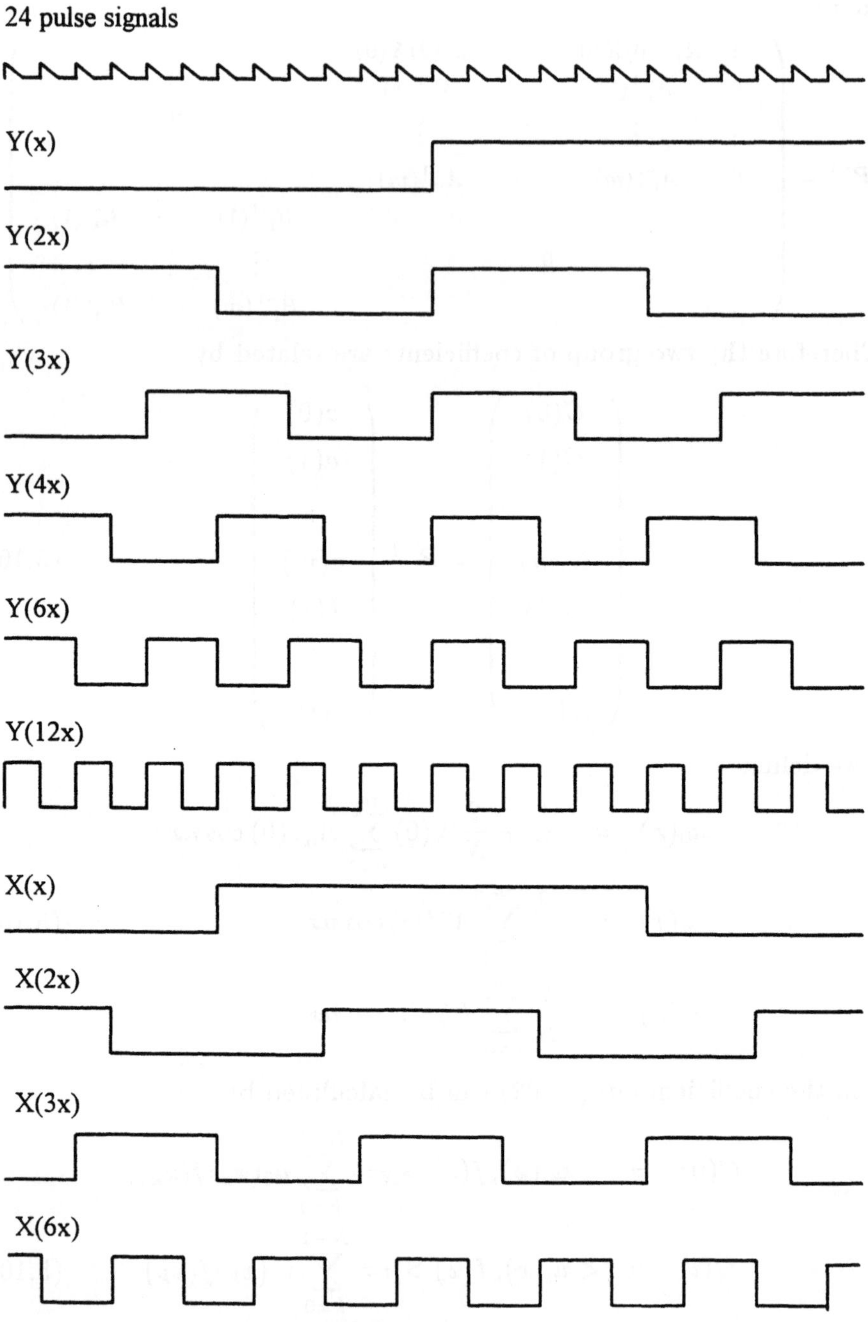

Figure 5.10: The square waves which can be obtained from 24 pulse signals by means of counters.

where $i = 1, 2, \cdots, m$.

Clearly this discrete transform is a generalization of discrete Fourier transform (DFT), which is suitable for digital signal processing.

5.7 Techniques of Common Waveform Analysis

In this section, we shall introduce the technical aspect of common waveform analysis, including generation of common waveforms, synthesizer and analyzer for common waveforms, filtering and multiplexing based on common waveforms, transformer from one kind of coefficients to another, and Fourier synthesizer by means of common waveforms.

1. **Generation of common waveforms**

Generally, square wave, triangular wave and trapezoidal wave are easily-generated functions in modern electronics.

As an example, it's quite convenient to obtain a system of square waves with different frequencies from a high frequency pulse by means of counting circuit. Concretely speaking, suppose that there are 24 pulse signals distributed uniformly in $(-\pi, \pi)$. By a $\div 12$ counter, the odd square wave $Y(x)$ can be obtained, and by setting the initial number of the counter be 6, one can obtain the even square wave $X(x)$. $Y(2x), Y(3x), Y(6x), Y(12x)$, and $X(2x), X(3x), X(6x)$ can be obtained in the similar way. See Figure 5.10.

Obviously it is somewhat hard to generate the sine-cosine system in the same way.

2. **Synthesizer based on common waveforms**

In practical applications by selecting proper relative magnitudes of enough common waveforms with different frequencies one can synthesize any signal at a certain accuracy with a summation circuit. See Eq.(5.55) and Figure 5.11. This is called a signal synthesizer based on common waveforms, or simply, common waveform synthesizer.

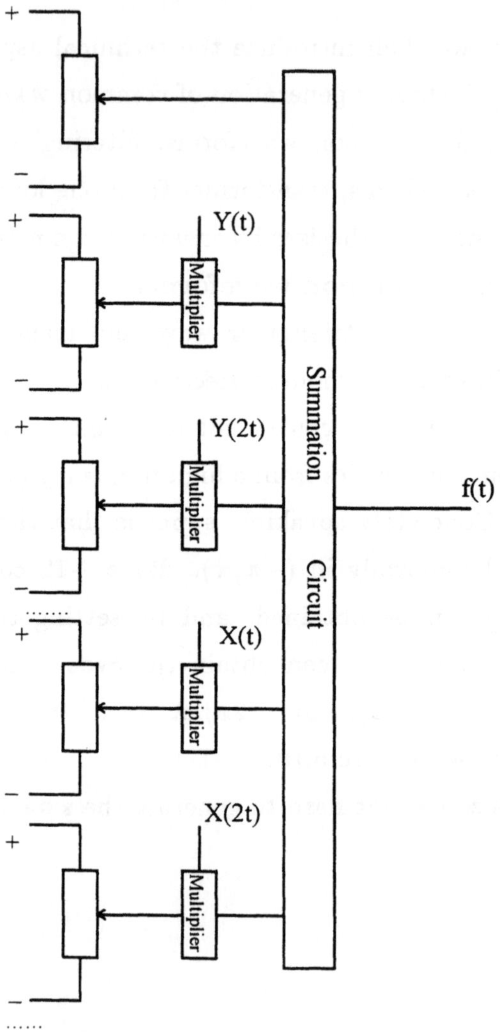

Figure 5.11: Schematic diagram of synthesizing any time function $f(t)$ by a system of common waveforms with different frequencies.

3. Analyzer for common waveforms

With the biorthogonal functions of common waveforms, one can decompose any signal into common waveforms by a computer program or by a specially designed device. The device is called a signal analyzer for common waveforms, or simply common waveform analyzer. See Eq.(5.58) and Figure 5.12.

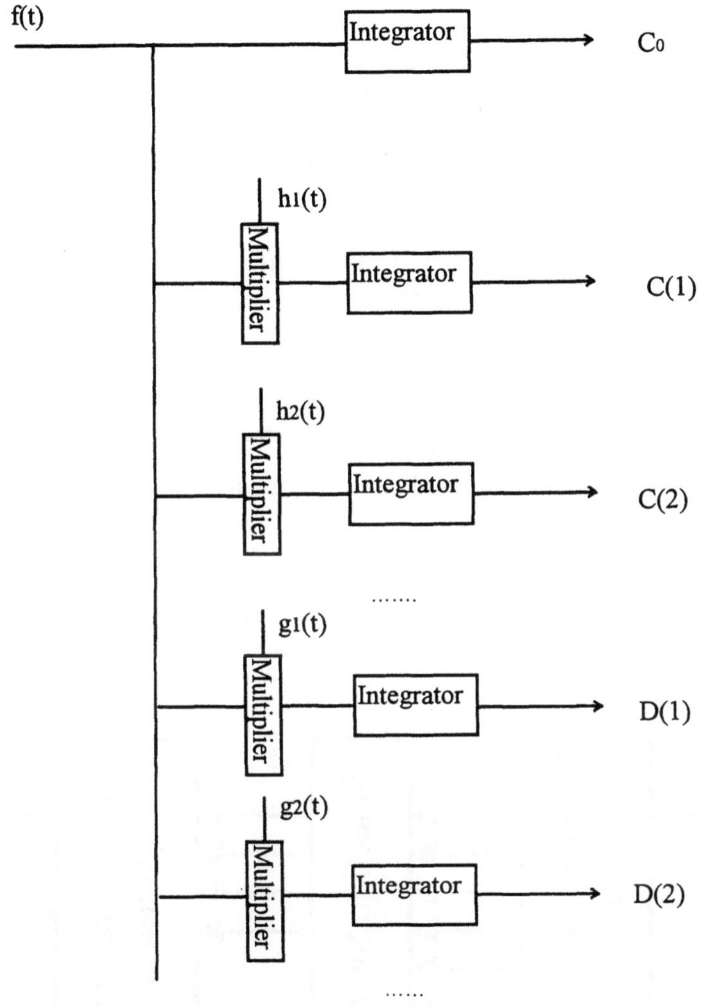

Figure 5.12: Schematic diagram of decomposing a time function $f(t)$ into a series of common waveforms with different frequencies by means of their biorthogonal functions.

4. Filtering based on common waveforms

If we decompose a signal into many harmonic components first and then recover the signal by a certain part of these components, Filtering based on sine-cosine functions can be carried out. Filtering based on common waveforms is very similar. There are three cases:

1) For a periodic analog signal $S(t)$, by decomposing a signal into a series of common waveforms with an common waveform analyzer and then recovering the signal with a certain part of these components, filtering can be carried out. In fact we have

$$\begin{aligned} S(t) &= C_0 + \sum_{n=1}^{\infty}(C(n)X(nt) + D(n)Y(nt)), \\ S_l(t) &= C_0 + \sum_{n=1}^{M}(C(n)X(nt) + D(n)Y(nt)), \\ S_h(t) &= \sum_{n=M+1}^{\infty}(C(n)X(nt) + D(n)Y(nt)), \end{aligned}$$

where $S(t)$ is the original signal, $S_l(t)$ the low frequency signal, and $S_h(t)$ the high frequency signal. See Figure 5.13.

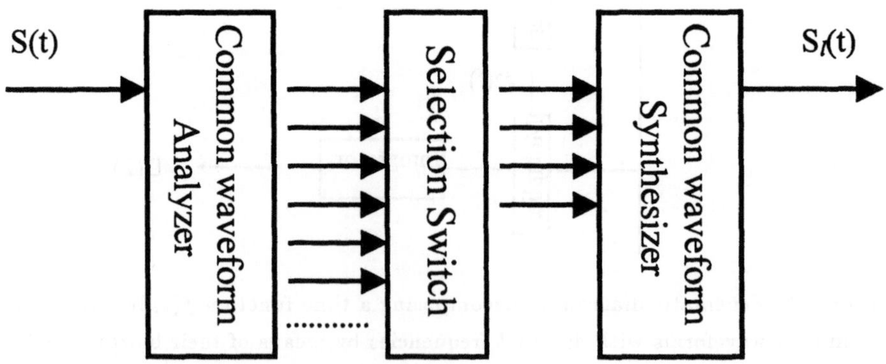

Figure 5.13: Schematic diagram of filtering based on common waveforms.

2) For a nonperiodic analog signal $S(t)$, we carry out common waveform transform first:

$$S(t) = \int_{-\infty}^{\infty} C(\omega)X(\omega t) + D(\omega)Y(\omega t)d\omega,$$

where

$$\begin{aligned} C(\omega) &= \int_{-\infty}^{\infty} f(t)\widetilde{X}(\omega t)dx, \\ D(\omega) &= \int_{-\infty}^{\infty} f(t)\widetilde{Y}(\omega t)dx, \end{aligned}$$

and then we can get low frequency signal and high frequency signal

$$S_l(t) = \int_{-M}^{M} C(\omega)X(\omega t) + D(\omega)Y(\omega t)d\omega, \tag{5.109}$$

$$S_h(t) = \left(\int_{-\infty}^{-M} + \int_{M}^{\infty}\right) C(\omega)X(\omega t) + D(\omega)Y(\omega t)d\omega. \tag{5.110}$$

3) For a signal with N discrete values, we carry out discrete common waveform first:

$$(S(t))_N = C(0)(1)_N + \sum_{n=1}^{m} C(n)(X(nt))_N + D(n)(Y(nt))_N,$$

and then we can get low frequency signal and high frequency signal

$$\begin{aligned} (S_l(t))_N &= C(0)(1)_N + \sum_{n=1}^{M} C(n)(X(nt))_N + D(n)(Y(nt))_N, \\ (S_h(t))_N &= \sum_{n=M+1}^{m} C(n)(X(nt))_N + D(n)(Y(nt))_N. \end{aligned}$$

5. **time-division multiplexing**

Since a group of common waveform coefficients is associated with a signal lasting a time interval, by transmitting common waveform coefficients quickly instead of the signal itself one can carry out time-division multiplexing of many signals. See Figure 5.14.

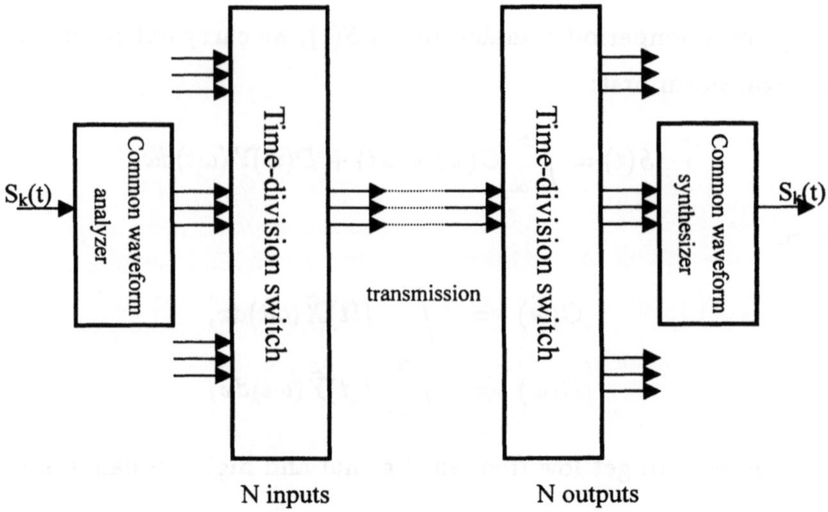

Figure 5.14: Schematic diagram of time-division multiplexing based on common waveform coefficients of signals.

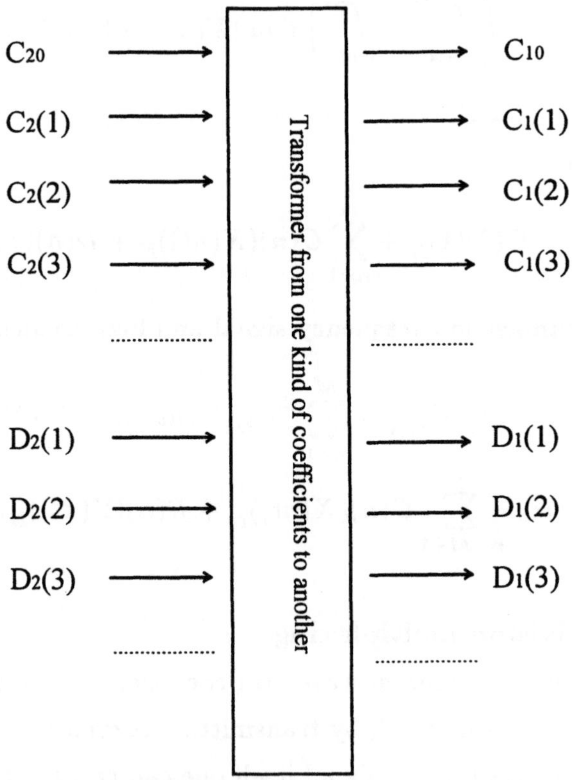

Figure 5.15: Transformer from one kind of coefficients to another.

6. **Transformer from one kind of coefficients to another**

A signal can be expressed as a square wave series, a triangular wave series, a trapezoidal wave series and so on. By the relation between them, one can turn one kind of coefficients (such as square wave coefficients) into another (such as triangular wave coefficients). See Eq.(5.64)-(5.65) and Figure 5.15.

7. **Fourier synthesizer by means of common waveforms**

It's more convenient to generate common waveforms with different frequencies than sine-cosine functions with different frequencies. Therefore when we synthesize a signal by means of its Fourier coefficients, we can first transform its Fourier coefficients into common waveform coefficients (such as square wave coefficients) by the relations (5.56), and then synthesize the signal with an common waveform synthesizer (such as square wave synthesizer). See Figure 5.16.

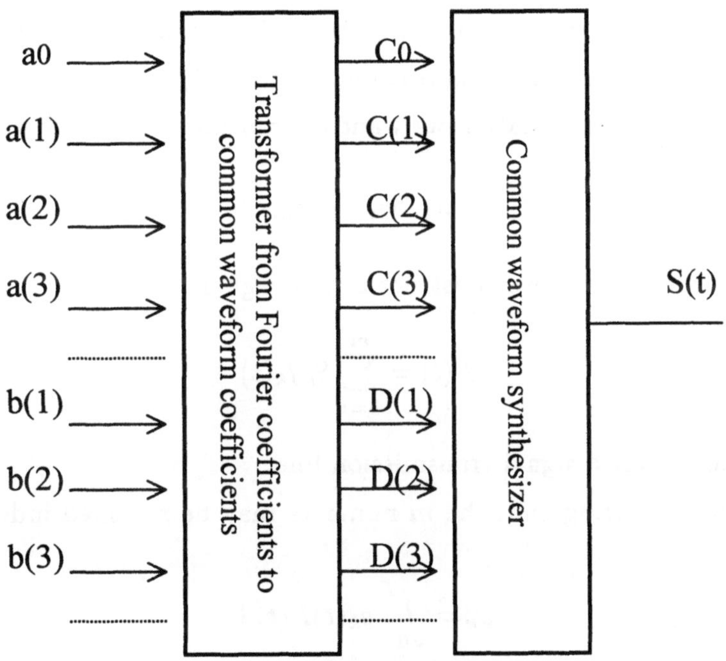

Figure 5.16: Schematic diagram of Fourier synthesizer by means of common waveforms.

8. Multiplexing based on biorthogonal functions

Multiplexing based on orthogonal functions, such as sine-cosine functions and Walsh functions has been considered carefully by H. F. Harmuth *et al* [5] [10]. Multiplexing based on common waveforms and associated biorthogonal functions is its generalization.

Mathematically, the function systems

$$f_1(t), f_2(t), \cdots, f_n(t), \cdots$$

and

$$g_1(t), g_2(t), \cdots, g_n(t), \cdots$$

are called biorthogonal in interval $[0, T]$, if we have

$$\int_0^T f_m(t) g_n(t) dt = \delta_{mn}$$

for any two integers m, n. Here δ_{mn} is the Kronecker's delta symbol. In other words, the function $f_n(t)$ is related to $g_n(t)$ only.

Suppose the information we want to transmit is m numbers:

$$S_1, S_2, \cdots, S_m.$$

At transmitting end, we construct a time signal:

$$F(t) = \sum_{k=1}^{m} S_k f_k(t)$$

and send it over a signal transmission line.

At the receiving end, the m numbers may be regained individually by

$$S_k = \int_0^T g_k(t) F(t) dt.$$

If we use common waveforms

$$X(t), Y(t), X(2t), Y(2t), \cdots, X(mt), Y(mt),$$

as carriers and associated biorthogonal functions

$$h_1(t), g_1(t), h_2(t), g_2(t), \cdots, h_m(t), g_m(t),$$

as demodulating functions, information can be transmitted by means of any waveforms rather than orthogonal functions only. See Figure 5.17.

If the time signals

$$S_1(t), S_2(t), \cdots, S_m(t)$$

change so slowly that they can be considered as a constant during the time T, this method is available as well. The only difference is that integrators are replaced by low-pass filters [5]. Figure 5.17.

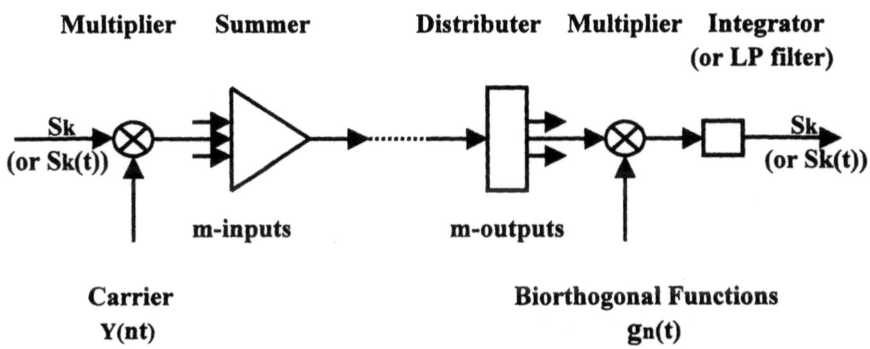

Figure 5.17: Schematic diagram of multiplexing based on common waveforms and their biorthogonal functions.

Though today we can generate a system of common waveforms with different frequencies in technique, we do not know the relations between signals and common waveforms, so that applications of common waveforms are limited seriously.

SUMMARY

From the above, we conclude that a function with period 2π can be expressed as a frequency series based on common waveforms, a nonperiodic function can be expressed as a frequency integral based on common waveforms, and a function known at only some points can be approximately expressed as a superposition of finite common waveforms. A series of techniques based on sine-cosine functions can be translated into techniques based on common waveforms. As a new and practical generalization of Fourier analysis, the common waveform analysis will be important in both theory and applications [11].

Bibliography

[1] L. K. Hua, Introduction to Number theory, (Springer-Verlag, New York, 1982). Chapter 1, Chapter 6.

[2] T. M. Apostol, Introduction to Analytic Number Theory, (Springer-Verlag, New York, 1976). Chapter 2.

[3] I. N. Bronshtein and K.A. Semendyayev, Handbook of Mathematics, English translation, 3rd edition, (Van Nostrand Reinhold Company, New York, 1985). pp. 574-579,747.

[4] H. F. Harmuth, Sequency Theory: Foundations and Application. (Academic press, New York, 1977).

[5] K. G. Beauchamp, Walsh Functions and their Applications, (Academic press, New York, 1975). Chapter 1.

[6] D. W. Tufts and G. Sadasiv, "The Arithmetic Fourier Transform," IEEE ASSP Mag., pp.13-17, Jan. 1988.

[7] C. Chui, An Introduction to Wavelets, (Academic, New York, 1992). Chapter 2.

[8] R. N. Bracewell, "Discrete Hartley transform, " J. Opt. Soc. Am. **73**,1832-1835 (1983).

[9] G. H. Hardy and E.M.Wright, An Introduction to the Theory of Numbers, 5th edition, (Oxford University Press, Oxford, 1973). pp.237.

[10] H. F. Harmuth, Transmission of Information by Orthogonal Functions, (Springer, New York, 1972). Chapter 3. pp 154-172.

[11] Y. Wei, Easily-generated Function Analysis, PhD thesis, Beijing University of Science and Technology, 1998.

INDEX

Arithmetical functions 2,114
Basis transform 54
Common waveform 118
Common waveform series 126
Common waveform transform 135
Completely multiplicative 4,115
Coordinate transform 54
Dirichlet inverse 6,116
Dirichlet multiplication 6,115
Divisor function $d(n)$ 3
Filtering 148
I-sawtooth wave 87,118
I-trapezoidal wave 45,121
II-sawtooth wave 90
II-trapezoidal wave 46
Möbius function μ 3,115
multiplexing 149,152
Square wave 14,118
Square wave series 128
Square wave system 16
Square wave transform 137
Three-valued function 92
Triangular wave 44,120
WASCMFC function 42,98
WCMFC function 118
WQSCMFC function 85

MIX
Papier aus verantwortungsvollen Quellen
Paper from responsible sources
FSC® C105338

If you have any concerns about our products, you can contact us on
ProductSafety@springernature.com

In case Publisher is established outside the EU, the EU authorized representative is:
Springer Nature Customer Service Center GmbH
Europaplatz 3, 69115 Heidelberg, Germany

Printed by Libri Plureos GmbH
in Hamburg, Germany